Fairey Firefly

An Illustrated History

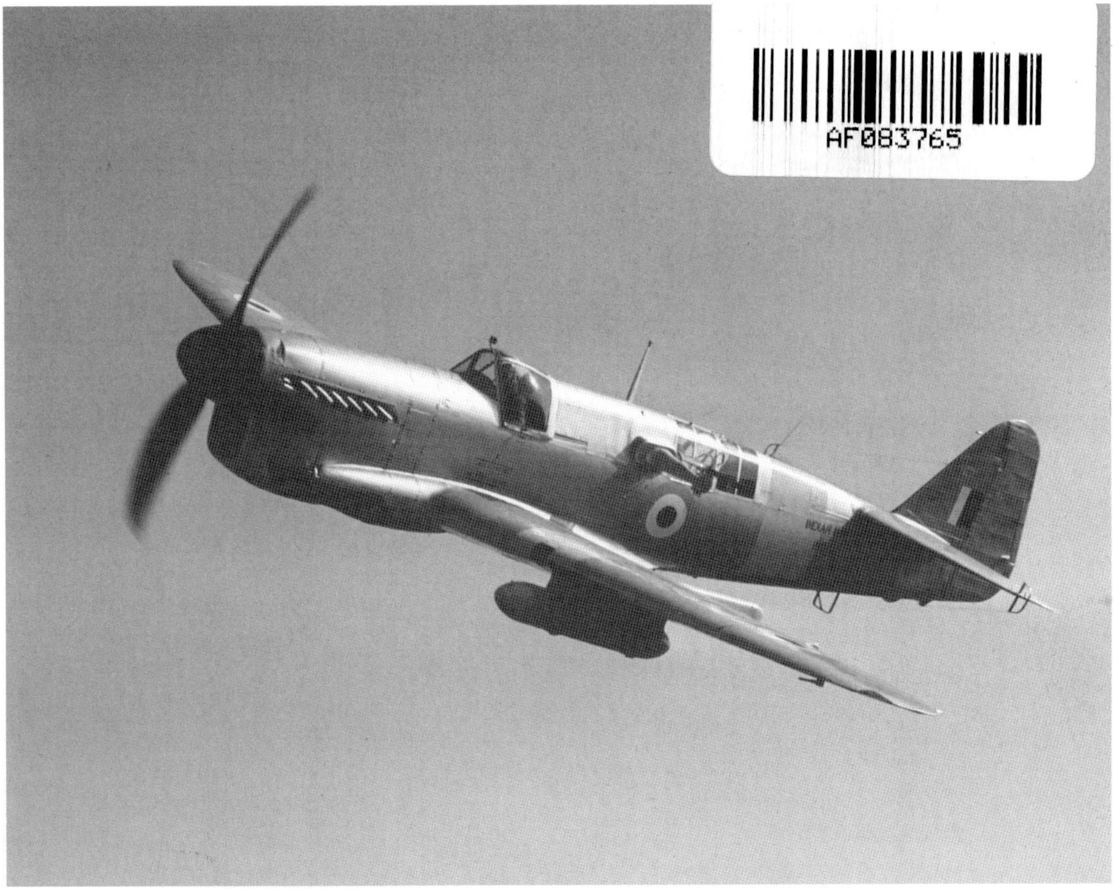

MATTHEW WILLIS

KEY Books

HISTORIC MILITARY AIRCRAFT SERIES, VOLUME 1

Contents page image: A Fairey Firefly Mk I, hook extended and Youngman flaps lowered, on its final approach to land on an aircraft carrier late in the Second World War.

Published by Key Books
An imprint of Key Publishing Ltd
PO Box 100
Stamford
Lincs PE19 1XQ

www.keypublishing.com

The right of Matthew Willis to be identified as the author of this book has been asserted in accordance with the Copyright, Designs and Patents Act 1988 Sections 77 and 78.

Copyright © Matthew Willis, 2020

ISBN 978 1 913295 89 9

20 21 22 23 24   10 9 8 7 6 5 4 3 2 1

All rights reserved. Reproduction in whole or in part in any form whatsoever or by any means is strictly prohibited without the prior permission of the Publisher.

Typeset by SJmagic DESIGN SERVICES, India.

Acknowledgements
The author would like to offer profound thanks to the following people who helped immensely with the book: Cheryl Baumgärtner, David Bull, Shell Bromley, Peter Menzies, Mary Ann Bennett, Coastlands Local History Group, Tony Jupp and Mark Moxley-Knapp.

Author's Note
All photographs are via the author unless otherwise stated.

Contents

Introduction 4
Chapter 1   The Firefly and the Second World War 7
Chapter 2   Post-war Fireflies 27
Chapter 3   Trainers 53
Chapter 4   Exports 61
Chapter 5   Korea 70
Chapter 6   Mk 7s and Drones 84
Chapter 7   The Firefly Close-up 88
Chapter 8   Preservation 93
Select Bibliography 96

Introduction

The Fairey Firefly two-seater strike-fighter emerged from a troubled gestation and birth to become one of the most widely used and effective aircraft of the Fleet Air Arm (FAA). It performed valuable service in the last year of the Second World War, with heavy armament, range and versatility contributing to its success. New models introduced after VJ Day improved the Firefly's performance and broadened its capabilities. The Firefly thus took part in a second major war, operating from light fleet aircraft carriers over Korea in the strike and reconnaissance roles. The Firefly made up part of the initial equipment of the new Dutch, Canadian and Australian Fleet Air Arms in the immediate post-war years and was operated by numerous other countries in first- and second-line roles. As naval aviation evolved, so did the Firefly with developments tailored to anti-submarine warfare, target towing and being used as a pilotless 'drone' for weapons tests.

The Firefly's development was lengthy and difficult, which reflected shifts in the nature of naval aviation and wartime design and production challenges more than the design itself. The first seeds of its creation were sown when the Admiralty and Air Ministry issued two requirements for new FAA fighter aircraft in 1938: specifications N.8/39 and N.9/39. These specifications were for two related and complementary two-seater fighters to replace the current Blackburn Skua and Roc. Like those aircraft, the replacements were to be a reconnaissance strike-fighter and a turret-fighter, based on the same basic airframe design. The specifications were challenging to meet, in particular requiring a very narrow width for hangar stowage and a very slow landing speed by the standards of contemporary monoplanes.

Marcel Lobelle, chief designer of the Fairey Aviation Company, prepared a submission to both specifications in late 1939 ready for the design tender conference in early January 1940. These schemes were typical of Fairey aircraft during this period and bore a general resemblance to the Battle bomber and the Fulmar two-seater naval fighter, the latter of which was due to enter service later in 1940. However, by this time the Admiralty had several months of experience of war to draw upon which reflected that some of the pre-war preconceptions on which N.8/39 and N.9/39 had been predicated were not correct. Chief among these was that performance was, in fact, very important. This may sound obvious, but in the late 1930s, the Admiralty and Air Ministry did not consider that carrier-based aircraft would have to confront high-performance land-based aircraft and consequently put more emphasis on versatility and ease of deck-landing.

Sure enough, on 23 December 1939, revised specifications were issued to the companies that had expressed interest. The updated specifications had dropped the turret-fighter and instead asked for a single-seater and a two-seater fixed-gun fighter, preferably based on the same basic design.

By this time, Lobelle had left Fairey and Herbert Eugene 'Charlie' Chaplin had replaced him as Chief Designer. Chaplin decided to start afresh and his new designs bore little in common with the existing ones. He introduced a semi-elliptical wing, somewhat reminiscent of the Spitfire's though simpler in shape, with a straight leading edge. Chaplin was focused on performance above all else and aimed to keep compromises from the accommodation of a second crew member and associated equipment to a minimum. The new design was aerodynamically clean with a slender, tapering rear fuselage. Fairey proposed a Rolls-Royce Griffon engine to power the aircraft, though it also offered a secondary scheme with a more powerful Napier Sabre.

The balance of high performance with acceptable deck-landing characteristics was a challenge for designers. Fairey chose to meet it through use of a device the company had recently experimented with: the Youngman flap. This helped to reduce take-off and landing speeds, and also offered a means of increasing endurance with a 'cruise' setting that increased wing area. This consequently reduced wing loading which also promised to reduce turn radius in combat.

In February 1940, H. N. Morrison, the Assistant Secretary of the Admiralty, wrote to the Air Ministry recommending that 200 of Fairey's revised N.8/39 design be ordered 'without going through the usual prototype procedure', as 'Messrs. Fairey's design could confidently be recommended as likely to prove satisfactory for an order "off the drawing board"'. By this time, the Fairey was promising a better performance than the original specification called for; it was expected to have a top speed of around 310 knots (357mph) which was 35 knots (40mph) faster than specified. (Even then, the Admiralty realised that the aircraft would not be fast enough for an interceptor if the performance of Axis bombers continued to improve.) The Air Ministry acquiesced and an order was placed for 200 airframes and 240 engines in March. A new specification N.5/40 was raised in May to cover the development.

Interestingly, despite the fact that a single-seater had also been asked for, the order was for two-seaters only. The Fifth Sea Lord remarked on 21 June 1940 that 'Lengthy consideration has been given in the light of war experience to the most suitable type of fighter aircraft for the FAA, and the pre-war conclusion has been confirmed that for normal and general functions of FAA fighters, the two-seater type should be retained in preference to the single-seater alternative'. At this point, the aircraft was expected to have a top speed of 360mph and to begin deliveries in 18 months, i.e. by the end of 1941.

Chaplin had noted in the submission that he had taken something of a gamble on the space required for a second crew member, lacking the time to produce a mock-up. Later, Commander M. S. Slattery, the Chief Naval Representative at the Ministry of Aircraft Production, said that 'To begin with, Faireys were asked to mock up a single-seater fighter with a hole cut in the back for the navigator but since then it has been fitted with every sort of equipment'. The original design evidently proved optimistic as the drawing produced for the design conference in March 1940 showed a distinctly deeper rear fuselage. The final design when it was frozen shortly afterwards had increased in length too with the observer's cockpit further aft and the engine moved forward to compensate for the shift in centre of gravity. In fact, the rear fuselage displayed a distinct 'hump' over its length, thus demonstrating a need to increase the observer's headroom and space for equipment.

In December 1940, the name 'Firefly' was allocated. This was not in keeping with the existing FAA naming policy, in which fighters were given the names of marine birds of prey and strike aircraft were named after predatory fish. It is unclear why the Admiralty chose to break with the existing policy, especially as it continued to be used with American aircraft, such as the Grumman TBF ('Tarpon') and F6F ('Gannet'). Fairey had already used the name Firefly for its series of biplane single-seater fighters from the 1920s to 1930s. However, the Firefly biplanes never served operationally with the UK services so Fairey considered it free to use. Interestingly, the Firefly's counterpart single-seater, based on Blackburn's N.8/39 design, also received a 'fire' themed name: the Firebrand.

Fairey Firefly

While the FAA struggled on with the Fairey Fulmar as its main fighter throughout 1941, Fairey geared up for Firefly production. The lack of 'the usual prototype procedure' did not mean that there would be no prototypes, but rather that the manufacturer would begin tooling up for production straight away. The first four aircraft on the production order would be hand built and would serve as prototypes. If testing threw up a requirement for more than small changes, then production could be held up, or shortcomings in the aircraft would have to be accepted. The original plan had been for the Firefly to follow on from the Fulmar at Fairey's Stockport factory, but delays to the Fulmar meant that deliveries would continue well into 1942. As a result, the Firefly would be built at Hayes, following the Albacore torpedo bomber.

The first prototype Firefly, Z1826, was completed in December 1941. It made its first flight with Fairey's veteran Chief Test Pilot, Chris Staniland, at the controls on 22 December 1941. Delays had begun to mount up with the programme, and the first aircraft flew at the time the FAA had originally expected the first deliveries of operational Fireflies. In February 1942, a 'whistle-blower' contacted the First Lord of the Admiralty to raise a string of management shortcomings within Fairey that were causing serious problems with the Barracuda and Firefly programmes. Rolls-Royce then experienced problems with the Griffon engine and the myriad engine spares and tools that would be needed to operate the Firefly in service. Despite these various woes, with the first Firefly in flight and production gearing up, it seemed that the FAA could soon look forward to receiving an aircraft it desperately needed.

A Firefly Mk I approaching the deck of a light fleet carrier in 1945. The approach seems good, but the batsman, with one bat extended and the other behind his back, is signalling 'too fast'.

Chapter 1
The Firefly and the Second World War

Below: The first prototype Fairey Firefly, Z1826, as it appeared in May 1942, shortly after Chris Staniland delivered it to Boscombe Down for the Aeroplane and Armament Experimental Establishment (A&AEE) to carry out brief performance and handling tests. Z1826 was first flown in bare metal but after the first few flights gained a representative FAA camouflage pattern. It seems unlikely that the underside was painted yellow, as was customary for prototype aircraft, as the tone of the yellow 'P' marking is noticeably different from the underside.

Dummy cannon barrels have been fitted to approximate the aerodynamic form of the guns. The windscreen is of the early, shallow type (later replaced by a significantly taller unit). The most obvious feature, however, is that the aircraft is lacking the sliding canopy hood. Z1826 undertook its manufacturer's trials between January and April 1942 with Staniland doing the lion's share of the work, supported by Foster H. 'Dicky' Dixon. Various improvements and adjustments were applied to the aircraft over this period, mostly as a result of Staniland's suggestions, and on 28 April he delivered Z1826 to the A&AEE. Unfortunately, during the flight from Great West Aerodrome to Boscombe Down, the hood came adrift and was lost. No replacement was available, so the A&AEE made the unusual decision, in light of the pressing need for the aircraft to progress, to carry out the programme with the cockpit open.

In hindsight, this was probably lucky for the A&AEE test pilots as it turned out there was a design flaw that meant the canopy had a tendency to detach, with potentially lethal consequences. Legendary test pilot Captain Eric 'Winkle' Brown unwittingly discovered the problem, as he explained to the author:

I was very much involved with the hood problems with the Mk I – the hood coming off. It killed two of their test pilots, Seth-Smith and Colin Evans. They were both killed with the hoods coming off – and then mine came off at Crail, severed the main spar of the tailplane, and lodged in the tailplane. I was able to get it back, and that's how we found out what the problem was. It was a design fault. It happened that we got away with it, because it had proved fatal, and the cause wasn't previously known. It was then coming off, and instead of flying away, it was dipping down into the cockpit and the front part of the hood, the curved part, was striking the pilot on the head.

Opposite: Two photographs of a prototype or early production Firefly Mk I released shortly after the type had entered operational service, following a long and troubled development. The aircraft has the configuration of the earliest machines with unfaired barrels on its four 20mm cannon and the early, shallow form of the windscreen and pilot's canopy.

The method of folding the Firefly's wings is well demonstrated in both images. The mechanism emulated Blackburn practice and represented a break from the Fairey style of wing folding employed on the Fulmar and Barracuda. On the Firefly, the wing hinged about the rear spar and twisted as it folded back to lie upright alongside the fuselage. This made for compact stowage and maximised hangar space.

The initial testing of the Firefly went well despite the unfortunate loss of the cockpit hood. The A&AEE highlighted minor issues with the controls, such as a tendency for the elevators to 'snatch' and the ailerons to overbalance, which would be relatively easy to address with changes to the control surfaces and circuits. The test pilots considered that 'If these controls can be satisfactorily modified, the aircraft should make a satisfactory naval fighter', adding that 'despite its weight, the aircraft possesses the characteristics and manoeuvrability of a smaller and lighter aircraft'. Improvements were made and a further round of trials took place in May 1942 which indicated that progress had been made. The A&AEE made additional recommendations, and Z1826 went back to Fairey for more adjustments in early June. The programme was able to continue because the second prototype, Z1827, had just become available.

However, disaster struck before the month was out. On 26 June, Z1827, with Staniland at the controls, suffered a catastrophic accident. The following day, the 5th Sea Lord, Rear Admiral, Lumley Lyster, informed the Admiralty that 'an accident took place with the prototype Firefly yesterday, the 26th June, in which the pilot was killed and the aircraft itself is reduced to a mass of tangled wreckage', adding that the programme was 'in trouble' and that 'suspicion must now fall on the Firefly as a Production aircraft'. The Admiralty wrote to Winston Churchill noting that Staniland's death had robbed the country of one of its most experienced test pilots and that, with the aircraft completely smashed and no immediate signs of what had caused the accident, the entire programme would have to be stopped until the cause was established. The reason was discovered in dramatic fashion – it transpired that the fabric of the elevators had failed. When this happened to another test pilot, he bravely stayed with the aircraft despite limited control to bring back the evidence.

Duncan Menzies, the lead test pilot at Fairey, Stockport, who had himself survived a horrific crash in a Fulmar a few months beforehand, described how the reason for Staniland's fatal crash was discovered:

> The circumstances of his last flight were not very clear at the time. I was not very fit having lost the empennage of a Fulmar at 400 indicated which produced a flick bunt. Colin Evans who was at Great West finally provided the evidence by belly landing a Firefly at Hatfield Bridge Flats some time later. A new form of stitch had been introduced to fix the fabric to the ribs of the elevators (and other parts). It was not a success and Chris must have lost rather more fore and aft control than Colin was left with.

Armed with this knowledge, Fairey could finally restart production. The opposite photographs were the first released to the public, accompanied by the following caption:

> Britain's latest carrier-based reconnaissance fighter, the Firefly, came off the secret list on Nov 3, with disclosure that the new plane is operating with the Fleet. While naval bombers attacked the *Tirpitz* in Norway recently, Fireflies engaged and shot down enemy planes which attempted to interfere. The Firefly's speed and other performance details were not revealed. The plane is equipped with special flaps to increase its manoeuvrability, permitting very tight turns, and its wings can be folded against the fuselage within 15 seconds.

Above and left: Two in-flight photographs of a prototype or early production Firefly from among the images released to the public after the aircraft's first operational missions. The first 19 Fireflies (including Z1827 which was lost after less than a month) were all employed on the development programme, with two more aircraft from the first 50. The first 70 aircraft were completed to the same standard as the aircraft in the photographs, though many of them were modified at various different stages, especially the development aircraft.

After the death of Chris Staniland, Dixon took over the lead on the programme. Work continued on the Firefly's controls, which were gradually improved. However, it took much time and many modifications before the aircraft was deemed fit for operational service and operation from an aircraft carrier. In January 1943, the Firefly was one of a handful of aircraft (including the Spitfire/Seafire and Lancaster) given top priority for production resources. This was despite the Admiralty delegation in Washington DC describing the type as 'unimpressive in 1943 and… completely outdated in 1945'. The Admiralty, however, considered that 'we think it is going to be an excellent night fighter, so good, in fact that the RAF have made enquiries about its use by them'. Their Lordships conceded that the aircraft's performance was low by the standards of contemporary day fighters; the earlier expectation of 310 knots (357mph) had not been met, and the best speed for the Mk I production variant was a pedestrian 274 knots (316mph). Nevertheless, the Admiralty hoped that performance could be improved and felt that the Firefly 'will be an extremely useful aircraft for armed reconnaissance'.

Finally, in July 1943 the Firefly was cleared for service. The lower photograph above was accompanied by the following caption:

BRITAIN'S NEW NAVAL PLANE – The Firefly, new British fighter–reconnaissance plane, wheels above a cloud in this official British photograph. The Firefly is fitted with four 20mm cannon and equipped with a camera for vertical photography.

A vital part of the Firefly's development programme was its ability to operate from a carrier deck. For this to be achieved, the aircraft needed excellent control at low speed, which the Firefly had previously failed to exhibit. According to Eric 'Winkle' Brown, the first deck trials were carried out in the Clyde with one of the early series aircraft, Z1829, on the rather wet and slippery looking deck of HMS *Illustrious*, from which this photograph was taken. Also present was the Firefly's single-seater counterpart, the Firebrand, which had first flown a couple of months after the Firefly. (The deck crew looking back at something behind Z1829 on its take-off run are actually looking at Firebrand prototype DD810, out of shot). The take-offs and landings were completed without mishap, but numerous further areas for improvement were identified. Notable on the photo is how low the cockpit enclosure is, with the pilot's head clearly visible above the top of the windscreen. The early-style hood is also visible – while this had a slightly curved roof, the sides were flat and a frame passed across each side, level with the pilot's eyeline.

The FAA's Service Trials Unit received a Firefly, Z1839, in May that year to assess whether the aircraft was fit for entry into operational service (including on aircraft carriers). By this time, multiple versions of the elevators and ailerons had been tried, as well as other alterations to the flying controls. A definitive configuration was established but not in time for the first 70 aircraft, which were all built to an older specification with a view to modification when the appropriate parts were available.

Winkle Brown was the service test pilot carrying out the deck landings associated with service trials. On first acquainting himself with the Firefly back in February, Winkle noted the position of the cockpit was slightly further aft than that of the Fulmar and 'coupled with some very obtrusive metal framing and decidedly restricted canopy headroom'. Considering Winkle's small physical stature, this brings home how constricted the cockpit must have been.

The second set of deck-landing trials took place aboard HMS *Illustrious* in June followed by a third set on the smaller trials carrier HMS *Pretoria Castle* in September, each time with Brown taking part. He noted that 'The Firefly handled well on the approach, but [the] view was seriously impaired by the previous windscreen and canopy faults', and that 'The touchdown itself had a nice, solid feel to it and once on the ground/deck the Firefly ran straight and steady'. He considered it 'a good average' as a deck-landing aircraft.

However, on Winkle's third take-off the canopy detached and caught on the tail as mentioned above. The Firefly was not out of the woods yet, but Winkle skilfully recovered Z1839 to Crail, and the incident allowed the design flaw with the original canopy hood to be discovered.

Above and left: Two photographs of a definitive Firefly F Mk I in flight. In October 1943, the Firefly was finally cleared for operational service. The designation of the first operational variant of the Firefly was given the prefix 'F' to signify 'fighter', according to the role prefix system applied from 1942; effectively this meant 'day fighter', as night fighters were distinguished by the prefix 'NF'. This was despite the Admiralty increasingly regarding the Firefly as a fast reconnaissance and light-strike type rather than a pure fighter.

One of the improvements that had allowed this to take place was the introduction of a new, taller windscreen and a redesigned canopy, which are particularly apparent in these photographs. In the first photograph, a Firefly F Mk I flies over the British countryside. The parallel frames of the new windscreen can be clearly seen as can the outward bulge of the canopy hood. The cannon barrels now have fairings covering all but the muzzle for improved aerodynamics.

The increased height of the new canopy can also be seen in the second photograph, in comparison with the images on pages 10–11. This photograph demonstrates the general form of the definitive Mk I, showing the semi-elliptical wing planform to good effect, and notably the Youngman flaps extended in the 'cruise' position. This allowed a tighter turn and also increased wing area for economical cruising. The Naval Air Fighting Development Unit tested Firefly Z1883 in late 1943 and considered that with the flaps in the cruise position, it could out-turn the Messerschmitt Bf 109 and Focke-Wulf Fw 190.

The Firefly entered service with 1770 Naval Air Squadron (NAS) at Yeovilton in October 1943. In reality, however, it would be 1944 before the squadron had its full complement of aircraft to the appropriate specification.

Right: Bomber's view: a Firefly Mk I breaks away after a mock-stern attack on a Handley-Page Hampden in February 1943.

Below: A Firefly Mk I on the deck of a fleet carrier in the Firth of Forth during the Second World War. The image demonstrates the wing-folding method (on a form pioneered by Blackburn) and the layout of the powerplant with all detachable cowling panels removed.

The first front-line Firefly squadron, 1770 NAS, moved to Scotland in early 1944, working up through the winter at Hatston, Orkney under Major Cheesman. In May they embarked on HMS *Indefatigable*, one of the new modified *Illustrious*-class carriers, to help bring squadron and ship to operational readiness. The main role for the Firefly was meant to be as an escort to *Indefatigable*'s Fairey Barracuda torpedo bombers. However, Major Cheesman pressed for the aircraft to have their own strike role. His wish was fulfilled with the addition to the Firefly's armoury of eight 3in rocket projectiles (RPs) as seen here. The early form of the RP installation required a heavy and drag-inducing rail and blast plate.

From April 1944, the FAA had been making a series of attacks on the Kriegsmarine battleship *Tirpitz* which was anchored in the Norwegian fjords and posing a threat to convoys to Russia. By the third attempt (Operation *Mascot*), 1770 Squadron was ready, and *Indefatigable* joined the older carriers *Furious* and *Formidable*. The Fireflies of 1770 Squadron had the important role of flak suppression, going in first to strafe the anti-aircraft emplacements on the ship herself and around the fjord. Since the first raid, defences had been strengthened considerably and several Fireflies were lost during *Mascot* and the following Operation *Goodwood* in August. Now blooded, the first Firefly squadron prepared to ship to the Far East.

Folding and unfolding the wings was accomplished manually and, as can be seen, required a sizeable team to manipulate the wings in and out of their folded position using long rods. It seems likely that the claims made to the press that the wings could be folded in 15 seconds were optimistic to say the least. The seagull on the flagstaff seems unimpressed.

Above, below and opposite: Before any Firefly squadron could become operational, each pilot had to spend hours perfecting carrier deck-landing. This did not always go smoothly, as this sequence of images from an incident aboard HMS *Theseus* (1945–46) attest.

In the above image, MT1D, an early Firefly Mk I from the first 70 machines with the low cockpit canopy, is moments from disaster. The hook has successfully caught a wire but the Firefly was not straight on landing, possibly touching down heavily on the starboard mainwheel which caused a swing to port. The Deck Landing Control Officer (DLCO) or batsman (often 'Bats' for short) is nowhere to be seen, most likely having thrown himself into the well behind his platform which the aircraft is about to run right over. Deck crew in the catwalk look on in horror as the aircraft barrels towards them.

In the image below the Firefly has pulled the arrestor wire taut but this has not stopped the port mainwheel dropping over the edge of the flight deck. A second or two later the Firefly has fallen into the port catwalk, as seen in the top opposite photograph, the still-spinning propeller throwing debris high in the air. The two larger pieces visible towards the top of the photograph are likely to be propeller blades themselves, as the plywood Rotol blades customarily sheared off on impact with a solid object. By the bottom opposite photograph, the Firefly is inverted and stuck in the port catwalk. It's unclear if the aircraft remained here long enough for the crew to escape.

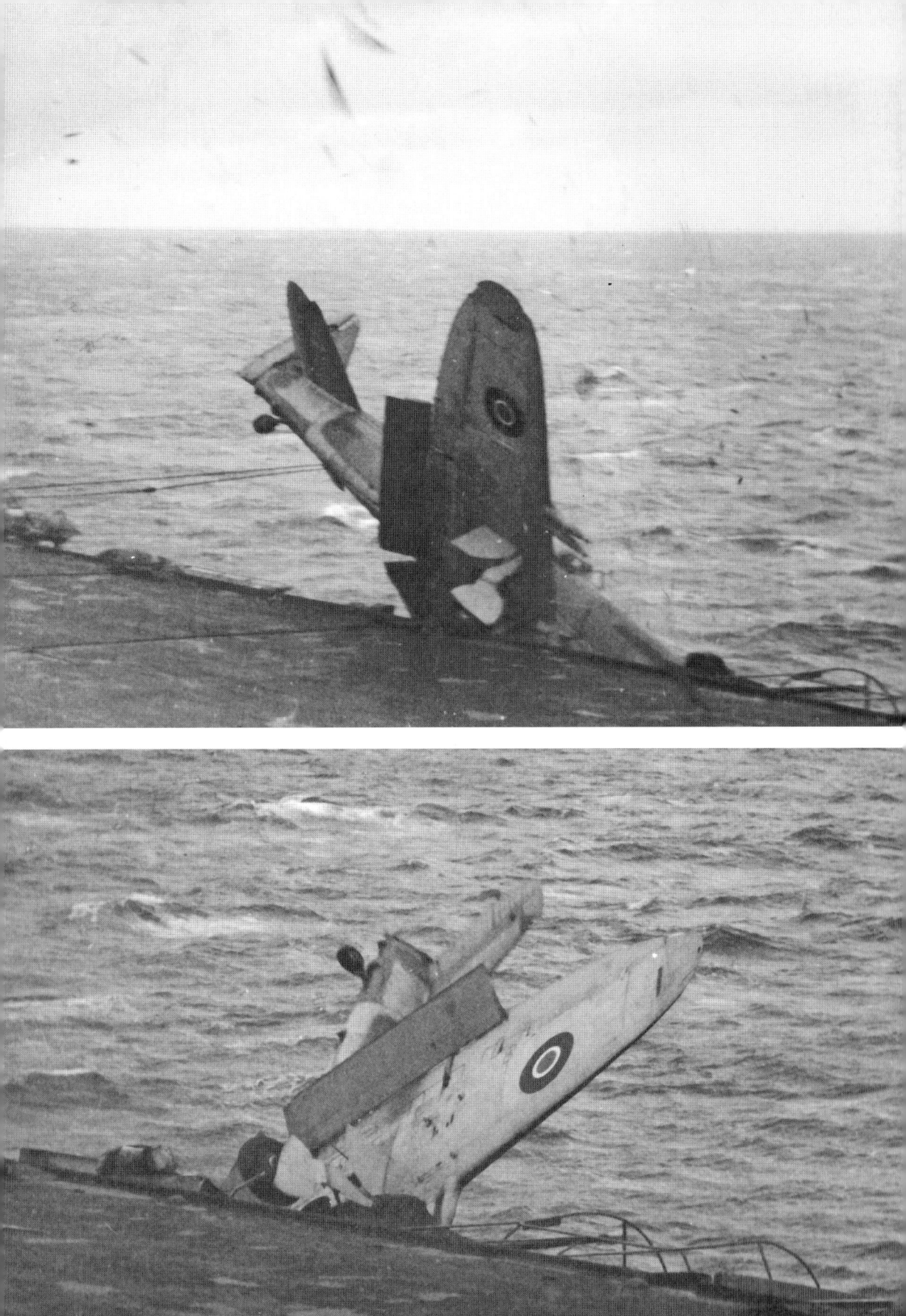

Left: Another, slightly less dramatic, deck-landing incident for a Firefly Mk I aboard HMS *Theseus* (one of the light fleet carriers that entered service at the end of the Second World War). Firefly IT1C, another early aircraft with the low windscreen and canopy, has probably missed all the wires and struck the crash barrier, nosing over and ending up sideways on the deck. The hook evidently did not fail to lower or pull out on catching the wire as the V-frame is visible between the fuselage and the wing. Winkle Brown suffered the former failure during a trial landing aboard HMS *Pretoria Castle*, the cockpit indicator light signalling that the hook was down when in fact it was not.

Unlike many Firefly barrier smashes, the damage does not seem too serious (incidents of this nature often triggered undercarriage collapses). It would represent something of a 'black' for the pilot in training who would nevertheless be expected to get into another aircraft and carry on with his training.

Opposite below: A Firefly FR Mk I touches down on a light fleet carrier as the batsman dramatically signals the 'cut'. The aircraft is in a tail-down attitude (essential for the hook to pick up a wire), though the tendency of the powerful Griffon engine to rotate the aircraft has led to one mainwheel touching the deck first, and the touchdown being somewhat right of centre. Both these things could be managed if the pilot was on his guard and didn't let a swing develop, but if allowed to accelerate it could result in one of the deck accidents shown in the images on pages 14 and 15. The Firefly's undercarriage, as with most British types of the era, was not as strong or as shock absorbing as on American designs, such as the Hellcat. Broken undercarriages were not an infrequent cause of deck accidents, though the Firefly could tolerate a small amount of asymmetric loading.

Below: Wave-off. The DLCO felt this Firefly Mk I was not going to be able to make a safe landing so instructed the pilot to abort. The procedure for a wave-off was to open the throttle to +9 boost, increase speed to 90 knots, retract the undercarriage and go round again. The power of the Griffon (1,730hp in the IIB variant, later upgraded to the Griffon XII offering 1,990hp) was such that it was sufficient to use climbing power rather than full take-off power. This meant that wave-offs were rather safer on the Firefly than on other types; Seafires could catch out an unwary pilot when the throttle was slammed right open to maximum boost, with the rapid increase in torque flipping the aircraft over.

This Firefly is an FR Mk I ('FR' for fighter–reconnaissance), the second version of the aircraft to enter operational service. This subtype is distinguished from the F Mk I by the pod under the nose, clearly visible in this shot. This contained a small, lightweight Air-Surface Homing (ASH) radar developed in the US (where it was known as the AN/APS-4). The ASH radar was supplied with its own fibreglass casing which resisted aerodynamic loads through being pressurised (this was simply achieved via a bicycle pump valve on the nosecone!). The self-contained scanner pod was intended to be mounted below an aircraft's wing (indeed, it is the size and shape of a small bomb) but the best position found for it on the Firefly was beneath the nose.

A Firefly FR Mk I, the second variant of the Firefly to appear after the F I and in most respects very similar. The ASH radar pod was carried on a boom, mounted below the forward fuselage at a slightly downward angle to the aircraft's axis. This allowed the radar the best 'view' forwards, downwards and upwards without compromising the aircraft's aerodynamics too significantly. This Firefly, named 'ANN', has the boom fitted with all the wiring and connections for the radar but lacks the self-contained nacelle with the scanner itself. The scope and other equipment for operating the device was fitted in the rear cockpit. At this stage, the radar still required a separate operator, being too complex for the pilot to operate as well as flying the aircraft, though later versions would make operation simpler so it could be used in single-seaters. The ASH radar was, at this point, primarily an anti-ship device which was intended to serve the same function as the earlier Air-to-Surface Vessel (ASV) radar (which was slightly more precise but much heavier and bulkier, and required drag-inducing 'Yagi' aerials). It would enable the FR I to locate surface vessels in conditions of poor visibility or at distances (up to 100 miles) beyond visual range.

The ASH installation was tested on the early series Firefly Z1970 at A&AEE Boscombe Down in early 1945. Its serial gained the suffix '/G' indicating that it was to be kept under guard whenever it was not being flown, such was the secrecy surrounding the radar equipment. It was found that handling was largely unaffected, though longitudinal stability (which was something of a problem from the outset) was somewhat degraded. In any event, handling remained good enough that FR I was cleared for service shortly after the A&AEE tests.

Above and below: An example of the Firefly night fighter (the NF Mk II Z1875) in February 1944. An increasingly important role for naval fighters in the latter part of the Second World War was carrying out night interceptions. German maritime patrol bombers, the Focke-Wulf Fw 200 and Junkers Ju 290, had been equipped with radar and were able to shadow convoys at night. The first FAA night fighter equipped with radar was the Fulmar NF Mk II, but the performance and reliability of this aircraft was so poor that a better platform was clearly needed. Early in 1943, the Admiralty discussed with Fairey the possibility of installing Airborne Interception (AI) Mk XI radar in a Firefly. The company undertook a study and concluded that the conversion was possible with a 15in extension to the nose to balance the weight of equipment in the rear cockpit. The antennae were mounted in nacelles on each wing root leading edge, the transmitter in one and the receiver in the other.

The Admiralty placed an order for 100 night fighters, but this would be taken from orders already placed. As a result, the Mk IIs had serials from the same block as early Mk Is and shared some similarities with them, such as unfaired cannon barrels and a low cockpit canopy. Firefly Mk I Z1831 was the first machine produced and underwent trials at A&AEE Boscombe Down in March 1943. These revealed severe handling problems and poor stability. Furthermore, the arrangement of the radar was problematic with difficulties in synchronising the two antennae and interference caused by the proximity of the scanners to the fuselage.

Development of the NF II continued and 37 production aircraft were built. Several served with 746 Squadron (the night fighter trials unit) but its fundamental problems were little closer to being resolved. In early 1944, the development of lightweight AI radar in the US promised a simpler solution. The APS-6 was based on the AN/APS-4 (ASH) radar being fitted to the FR I. As it turned out, the APS-6 was not made available to the FAA, but the standard ASH could be used in the airborne interception role. It just required simple adaptations to the Firefly FR I radar installation and straightforward changes, such as painting cockpit interior surfaces black to adapt an FR I into a night fighter. The 37 NF IIs completed were either converted back to Mk I status or scrapped, and the Firefly NF I went into production alongside the FR I; the only external difference between these two sub-variants was a flat strip fitted over the exhaust stubs to shield the pilot's eyes from the glare on the night fighter.

Firefly F I DT934 of 1770 Squadron about to take off from HMS *Indefatigable* on 4 January 1945, equipped with RPs to attack an oil refinery on Pangkalan Brandan, Sumatra.

In the spring and summer of 1944, the focus in the European war was on the Allied landings in France, though the FAA continued to operate in Norwegian waters against German shipping. The FAA's main task now shifted to the Far East where the Royal Navy (RN) was building its most powerful force of aircraft carriers of the war. After its baptism of fire, the first Firefly squadron, 1770 NAS, was dispatched east aboard HMS *Indefatigable*. They arrived at Ceylon on 10 January 1945, ready for raids on Japanese targets in the East Indies in the new year with the British Pacific Fleet (BPF). This new formation had been created from the Eastern Fleet in November 1944 and signalled the RN's ambitions to take an active part in the war against Japan.

A series of raids against Japanese oil refineries began in late 1944, partly because of their strategic value and partly to help forge the BPF into a combat-hardened force that would be able to hold its own alongside the US Navy. In this photograph, DT934 (which wore the code 4-K) is about to take part in Operation *Lentil*, which involved flying 130 nautical miles over Sumatra, with an 11,000ft mountain range to negotiate, before reaching the target. This was no easy feat for heavily loaded aircraft optimised for low-level operations, in hot conditions that sapped engine power. As with the *Tirpitz* raids, the Fireflies were to go in first, attacking defences with rockets and cannon fire. However, they caught the anti-aircraft batteries and defending aircraft unawares and although there was little flak and no aircraft preparing to take off, they strafed the airfield anyway. Some aircraft spotted a small tanker and attacked that, setting it on fire. On the return to the carrier, Major Cheesman was prevented from landing by a Seafire crash on *Indefatigable*'s deck and had to ditch his Firefly, though he and his observer, Lieutenant Wilkey, were recovered safely.

The photograph was taken by Lieutenant C. Trusler, the RN official photographer, and issued to the press alongside an Admiralty communique with the following caption:

For first publication weeklies, 9th February 1945.

On January 4th British carrier-borne aircraft of the Fleet Air Arm delivered a very accurate attack against the enemy oil refinery at Pangkalan Brandan in Sumatra. Weather conditions were excellent and the whole weight of bombs and missiles from the Avengers and Fireflies fell within the target area. The power-house and other important plants, together with oil tanks and buildings, received direct hits. This picture was taken onboard the British carrier *Indefatigable*, which took part in the attack.

Picture shows – a rocket-carrying Firefly awaiting the signal to take off from the flight deck of the *Indefatigable*.

The same Firefly of 1770 Squadron, DT934/4K, flies over the flagship HMS *Indomitable* with her two squadrons of Grumman Hellcats, 1839 and 1844, ranged to take off on Operation *Meridian* (the raids on Palembang).

Operation *Lentil* was judged a success, and was to be followed up with attacks on targets at Palembang (the Japanese HQ in Sumatra and home to two of the largest oil refineries). No fewer than four fleet carriers were taking part. Once again, the striking aircraft would have to fly over a mountain range en route to the targets, and the role of 1770 Squadron's Fireflies would be the same: to suppress defences for the following strike aircraft. Unfortunately, technical failures affected three Fireflies, leaving just nine for the raid. They had only just taken up their place ahead of the Avengers when Japanese fighters were spotted on an intercept course. In the confusion, most of the Firefly crews missed an order from the strike leader to strafe barrage balloons that were threatening to prevent the Avengers from bombing. The balloons proved resistant to the gunfire of those aircraft that did hear the order, leaving most of the Avengers to risk diving through the balloons. Then, acting as close escort, the few Fireflies did their best to keep the Tojos and Oscars off the Avengers' backs in the absence of the Corsairs and Hellcats, but heavy losses were incurred in return for around a third of the targets destroyed.

The fleet replenished at sea, then returned for the second *Meridian* raid (this time against the refinery at Songei Gerong). In light of the stiffer than expected defence, 1770 Squadron's Fireflies would be solely dedicated to close escort with no strafing role. Although, once again the strike leader directed them to attack barrage balloons when the strike force approached the target. Three balloons were dispatched, not enough to make the Avenger pilots' lives much easier. However, Fireflies did succeed in going toe-to-toe with Japanese fighters, using the Youngman flaps to good effect, turning with the manoeuvrable Oscars and Hamps and claiming three shot down. One Firefly was lost after having been seen engaging enemy fighters.

Operation *Meridian* had led to heavy losses and dealt a blow to morale among BPF aircrews. However, the mission had proved that the RN could carry out complex, co-ordinated multi-carrier strike missions, and the stage was now set for the force to move into the Pacific.

The air, deck and hangar crew of Firefly Mk I '4-C' ashore in the Far East. After Operation *Meridian*, the BPF sailed to Fremantle and then Sydney where the crews could rest and recuperate and make good the losses from the East Indies missions. On 10 February 1945, *Indefatigable* was nominated as part of Task Force 112 and prepared for operations in the Pacific under the command of the US 5th Fleet. The BPF was designated Task Force 57 (TF57) for the purpose of its operations in the Pacific. At the end of February, the fleet sailed from Sydney, joining the 5th Fleet on 18 March at Ulithi. As TF57, the BPF rotated with the US TF58 to carry out attacks on the Sakishima Gunto islands which were being used as a staging post for kamikaze missions. The resulting campaign was designated Operation *Iceberg*.

From 26 March, rocket-equipped Fireflies from 1770 Squadron took part in raids on the two main islands' airfields, hitting the flak emplacements while Avengers bombed the runways. In addition, they attacked harbour installations and any shipping they encountered as well as a radio transmitting station. While on station, TF57 was subjected to repeated kamikaze attacks. It was believed these originated from Formosa, so the US Admiral in command withdrew the British carriers from Sakishima Gunto and directed their attacks at the suspected kamikaze base. On 12 April, two Fireflies surprised a formation of five Japanese bombers heading for Okinawa and shot down all but one. At the beginning of June, *Indefatigable* returned to Australia.

Fireflies and Seafires from HMS *Implacable* in June 1945, including DK431/275 of 1771 NAS, named *Evelyn Tentions*, which, according to some sources, was the personal aircraft of Lieutenant Commander W. J. R. MacWhirter, the squadron's commanding officer. The aircraft wears the new style of markings adopted when the BPF moved into the Pacific. Instead of the two character codes worn previously, the aircraft is identified by a three-digit number beginning with the type's number of seats. The national markings have been updated from the Eastern Fleet's small blue and white roundels to the larger blue and white roundels with bars (similar to the markings worn by United States Navy (USN) aircraft except with a cockade in place of the star). In addition, all 1770 Squadron aircraft had a yellow map of the Japanese home islands painted on the cowling, and some aircraft had a name based on word play, one example being *Lucy Quipment*.

No. 1771 Squadron was the second Firefly unit to become operational and, like 1770 NAS, cut its teeth with strikes off Norway before heading east. HMS *Implacable* was the only RN carrier to see action in June 1945, having just arrived from the UK, while the existing carriers of the BPF were enjoying a rest and refit in Australia. Raids on Truk on 14 and 15 June helped sharpen the carrier's air group before hitting more demanding targets. She then met with the rest of the BPF (apart from *Indefatigable* which was delayed by mechanical faults) in July for raids on the Japanese home islands. By this time, 1770 Squadron had stood down after having been in action more or less constantly since the previous spring and having been replaced aboard *Indefatigable* by 1772 Squadron.

Above: A Firefly F I of 812 NAS skims over the round-down to land on HMS *Vengeance* in 1945. By the end of the war, there were three Firefly squadrons in the Pacific: 1770 (non-operational), 1771 and 1772. The first two were disbanded quickly and passed their aircraft to squadrons newly arrived in theatre.

Vengeance was one of the first four *Colossus*-class light fleet carriers (which together formed the 11th Aircraft Carrier Squadron (ACS)) due to join the BPF late in 1945. Some of the 11th ACS carriers and their squadrons remained in the area for up to two years afterwards helping to maintain the RN's presence in Pacific waters.

The original air groups of the 11th ACS carriers were made up of Corsairs and Barracudas but with their changing role, following the end of hostilities, the Barracuda squadrons were re-equipped with Fireflies. John Dickson, a pilot with 812 Squadron, told the author 'We came back down to Sydney at the beginning of January in 1946, and went off on leave for three weeks, which was well overdue by then – we didn't have any leave at the time of VJ Day. When we came back from leave, the Barracudas had all gone, thank goodness, and we took over some Fireflies. And we then continued working up with Fireflies, which was quite fun'.

Like most Fireflies in the Pacific, this aircraft is fitted with RP rails.

Opposite below: Firefly Mk Is from 827 Squadron aboard HMS *Triumph* in port at Trieste in 1948. Like *Theseus*, *Triumph* was commissioned in 1946 but took up her first posting with the Mediterranean Fleet, with which she served as the flagship of the fleet's Flag Officer, Air; 827 Squadron was part of the 13th Carrier Air Group (CAG), initially equipped with Firefly FR Is. In 1949, the squadron gave up four of these machines to accommodate four night fighter variant NF Is of 812 Squadron. In this period, several Firefly squadrons took on a section of night fighters which were known as 'Black Flights'.

These Firefly Mk Is still wear the wartime Temperate Sea Scheme camouflage. Interestingly, the second Firefly back on the starboard side has one older, wartime-style upper wing roundel.

This was a press photograph issued to the media in 1948, accompanied by the following caption:

TRIESTE – Looming high over the waterfront here, planes rest on the deck of HMS Triumph, 18,000-ton British carrier, during ship's visit prior to the crucial Italian elections. 17/4/48.

Above: Firefly F I DV117/286 narrowly avoiding disaster after landing on HMS *Venerable* in 1947. *Venerable* was another of the 11th ACS light fleet carriers that arrived in the Far East just as the war ended and, like 812 NAS, 814 Squadron had worked up as a Barracuda unit. Aircraft DV117 had belonged to 1771 Squadron in 1945 (as indicated by the lightning device on the nose) but it now wears the 'V' tail letter associated with HMS *Venerable*. Three deck crew from the port catwalk are busy freeing the tailhook from the arrestor wire while another two race from starboard to assist. The Firefly will have to be manhandled to a safer position then pushed forward to be parked in 'Fly 1', so the barrier can be raised and the next aircraft can land. Note the observer's cockpit with the panel open at the top.

Left: A cartoon of a Firefly FR I, part of a mural covering two walls of a barrack block in the Royal Naval Air Station at Dale, Pembrokeshire. The mural was created by Sub-Lieutenant (A) 'Dax' Dashfield RNVR, an engineering officer with 790 Squadron, whose art also appeared in the station's magazine. He always 'signed' his cartoons with a representation of a winged angel in a flying helmet and goggles. The mural is maintained by the Coastlands Local History Group, which believes that the mural is likely to have been painted in late 1947 or even 1948, shortly before the station was due to close, or it would surely not have survived – indeed, nor would 'Dax' have got away with defacing the King's property. (Photograph courtesy of the Coastlands Local History Group).

Below: Firefly Z1835, which started life as a Mk I but as seen here is a 'missing link' in the Firefly's evolution. After the failure of the Mk II night fighter, the next area for attention in Fairey's design offices was to achieve a step forward in the performance of the standard day fighter. This was of some urgency, considering the production Mk I had turned out to be a full 41mph slower than predicted. In 1941, H. E. Chaplin became aware of a projected Griffon development with a two-stage supercharger, the Griffon 61, which would be significantly more powerful than the IIB model used in the Mk I (and later see much use in the Spitfire Mk XIV). The Admiralty was keen for this development to be implemented and was prepared to make large production orders.

Rolls-Royce's original proposal to Fairey involved wing leading-edge radiators which were soon to become popular with the de Havilland Mosquito, Blackburn Firebrand and prototype Hawker Tempest and Fury aircraft, thus demonstrating their potentially superior characteristics with regard to overall drag reduction.

However, it appears these were not available and the test aircraft for the Mk III was cooled via a large, circular 'beard' radiator requiring an almost cylindrical cowling. This cumbersome-looking arrangement reintroduced all of the stability and control problems that had been painstakingly resolved in the early Mk I in a manner even more severe than before. There were further problems with cooling and engine fumes entering the cockpit. The Mk III did display some improvement in performance (it registered a top speed of 345mph at 12,000ft), although the handling and other problems meant that the Admiralty considered it was 'unsuitable for service use'. The aircraft was tested from May 1943 to May 1944 without significant improvements being found and, with other potential developments seeming more promising, the Mk III was dropped in December 1944.

Chapter 2
Post-war Fireflies

The Fairey Firefly Mk I was arguably obsolescent when it came into service, at least in its intended role as a fleet fighter. However, it had proved useful as a fast reconnaissance and light-strike aircraft, and in a dogfight it was certainly able to hold its own against high-performance adversaries. Perhaps more importantly, in the early post-war years there were few alternatives. The terms of the lend-lease agreement by which the FAA had acquired large numbers of Avengers, Corsairs and Hellcats, which were undoubtedly favoured by 1945, meant that it could not keep them. Most of those in the Far East were simply pushed off their aircraft carriers into the sea. The Firefly, being a British type, was available.

In fact, it was more suitable than some other types for the immediate post-war period. As noted above, Fireflies replaced Barracudas in the Pacific after the end of hostilities. As a reasonably good deck-landing type that could be used for long patrols and light strikes against sea or land targets, it was ideal for policing the peace. Fireflies, something of a rarity in FAA service during the last years of the Second World War, were, in the new peace, becoming ubiquitous.

Furthermore, Fairey had not rested on its laurels. After the failure of the Mk III programme, much more promising avenues had opened up. Placing the radiator in a duct below the rear fuselage (resembling a crude version of the North American Mustang's sophisticated 'Meredith' cooling configuration) realised no benefits when wind-tunnel tested. However, the wing leading-edge radiators that Rolls-Royce had mooted back in 1941 were considered again in 1944 for use with a Griffon 72 of 2,245hp, with wind-tunnel tests showing promising results. The first prototype, Z1826, was modified with an aerodynamic mock-up of the installation, and finally on 25 May 1945, a true prototype Z2118 appeared with wing leading-edge radiators and the original chin radiator finally banished. Production Firefly Mk IV aircraft started to become available in May 1946.

The designation of FAA aircraft changed from Roman numerals to Arabic in the post-war period, so henceforth, mark numbers will use that style.

The following sequence of photographs (see pages 28–32) are from HMS *Ocean*'s commission in the Mediterranean from 1946 to 1948. Although in 1945 and 1946 the RN was operating more fleet aircraft carriers than at any time in its history, the post-war drawdown was swift and by 1948 there were only two carriers in full commission with air groups embarked: *Triumph* and *Ocean* (both in the Mediterranean).

HMS *Ocean*'s air group consisted of 816 NAS, operating Firefly FR 1s, and 805 Squadron, which was nominally equipped with Seafire Mk 15s. However, for a period the Seafire squadron had to switch to Fireflies in order to allow its pilots to maintain their currency due to problems with the Seafire Mk 15; 805 operated its Fireflies as single-seaters. The pilots of 805 Squadron also took on night fighter duties from 816 during this period.

The image above shows Firefly PP557/O6G of 816 Squadron's night fighter 'Black Flight' during HMS *Ocean*'s spell with the Mediterranean Fleet. The aircraft was from the last order for Fireflies from which aircraft were delivered during the Second World War, although PP557 was not completed until after the end of hostilities. This photograph demonstrates several ways in which the Firefly Mk 1 had been developed from wartime specifications. The ASH radar gained a shroud, covering the boom mounting, cleaning up the aerodynamics and offering some protection from the elements. The cumbersome wartime RP launchers have given way to the much neater, lighter and less drag-inducing 'zero-length' pylons, after it was established that the blast plate and long rails were unnecessary.

PP557 also demonstrates the evolution of markings since the end of the Second World War. The overall colour scheme is the same – a disruptive camouflage of Extra Dark Sea Grey (a blueish dark grey) and Dark Slate Grey (a greenish mid grey) on upper surfaces and Sky (a pale grey-green) on under surfaces. Reflecting peacetime conditions, the aircraft's serial is repeated in extremely large characters under each wing, and roundels in all positions are of the 1942 red, white and blue design, with a thin white ring around the central red circle and a thin yellow ring around the outside of the fuselage roundels. In another typical peacetime touch, the squadron's badge is displayed on the engine cowling.

Above: A range of 18 Fireflies aboard HMS *Ocean* in the Mediterranean, 1946–47. The full establishment of 816 Squadron was initially 12 aircraft in July 1945, then four night fighters were added in May 1946. The presence of an 'extra' two aircraft in this photograph suggests that some are aircraft flown by 805 Squadron as single-seaters in the absence of their Seafires (which had been withdrawn as dangerous). The aircraft are wearing the earlier arrangement of markings with fuselage codes consisting of the number '6' and an individual aircraft letter, prefixed with 'O' to match *Ocean*'s deck letter.

Right: Firefly FR 1s of 816 Squadron in extremely close formation over Grand Harbour, Valletta, Malta, 1947–48. The style of markings has changed from the earlier photographs, with the carrier letter 'O' now moved to the tail and codes changed to a BPF-style three-digit arrangement.

Above: The same formation of 816 Squadron Firefly FR 1s, this time over the RN destroyer anchorage at St Paul's Bay at the northern end of Malta. Visible on aircraft 208 is the partially obliterated carrier letter 'O' (recently moved from the rear fuselage to the tail).

Below and opposite above: Firefly FR 1 PP546/216 coming to grief aboard HMS *Ocean*, 1947–48. The unfortunate PP546 evidently missed all the wires on landing, collected the first barrier, tipped up onto its nose and continued skidding towards the second barrier (the ASH radar pod having ripped off). In the first image, the Firefly has just stopped moving a few feet before it would have hit the second barrier but has caught fire. The deck crew leap into action with fire hoses. In the next photograph (see opposite above), taken a minute or two later, the Firefly has been sprayed with fire suppressant foam and any fire has evidently been extinguished quickly as there is no obvious fire damage to the aircraft.

Below: Parade on the flight deck of HMS *Ocean*, anchored off Sardinia, 1947–48. Two 816 Squadron Firefly FR 1s are ranged aft. Until 1948, much of *Ocean*'s work in the Mediterranean could be described as 'showing the flag' – ostensibly ceremonial, but with a strategic purpose to it.

The nascent Cold War between the US and the USSR and their allies became rather hot for the ships of the Mediterranean Fleet in 1946, when a series of confrontations known collectively as the Corfu Channel Incidents took place. Albanian artillery fired on RN warships in May that year, while in October the destroyers *Saumarez* and *Volage* struck mines when passing through the straits and were seriously damaged. The following month, *Ocean*'s Fireflies provided aerial cover to RN ships clearing mines from within Albanian waters (Operation *Retail*) in case of interference from Albanian or allied forces.

Left: Firefly PP542 taxies forward to park at 'Fly One' after landing, while the barrier is raised behind to allow the next aircraft to land.

In May 1948, *Ocean* joined the task force overseeing the withdrawal of British forces from Palestine, hers being the only aircraft present to cover the departure of the High Commissioner. On 15 May, the last day of the British Mandate in Palestine, *Ocean*'s aircraft provided a flypast as the High Commissioner left Haifa on Admiral Troubridge's barge.

Below: A dramatic image showing Firefly FR 1 PP540 sprayed with fire suppressant foam on deck aft. The machine appears to be ranged for take-off, so the fire may have been spontaneous.

This photograph shows the post-war installation of the ASH radar (with a fairing over the mounting boom) and the zero-length rocket pylons to good effect. The latter were attached to a plate that sat flush with the wing's under surface and created much less drag than the older installation.

Right: Firefly NF 1 night fighter PP468 taking off from HMS *Indomitable* in 1952. This aircraft was one of the first Fireflies to be completed after the end of the Second World War, and in 1950 was one of those that went to war from HMS *Triumph* in Korea (see Chapter 5).

Above and right: Firefly Mk 1 PP425 flying over the UK coast. This aircraft demonstrates the markings applied to Fireflies from 1947: Extra Dark Sea Grey upper surfaces and Sky under surfaces, with the later style red, white and blue roundels in a 1:2:3 ratio. It has the Second World War arrangement of its ASH radar pod with no aerodynamic shroud, as seen on the HMS *Ocean* aircraft above. PP425 appears here, typical of the NF Mk 1s of the late 1940s and early 1950s, with ASH radar and exhaust glare shields. In common with many Mk 1s, PP425 was converted to become a T Mk 3 observer trainer.

Above: Prototype Firefly Mk 4 Z2118 during its test programme in 1946. With the failure of the Mk 3 set out in the previous chapter, Fairey was able to rescue the Firefly with a cooling arrangement that was becoming increasingly popular in high-performance piston-engined aircraft: wing leading-edge radiators. In this layout, rather than being installed in a duct outside the airframe profile, the coolant radiators were kept within the frontal area of the wings. Leading-edge radiators allowed Fairey to introduce a more powerful Griffon, the 2,330hp model 72, while mitigating its increased cooling requirements.

Fairey modified a Mk 1, Z2118, to the engine and cooling arrangement of what was now being referred to as the Mk 4. The Mk 3 prototype, Z1835, was also modified to Mk 4 standard, and two more Mk 1s, MB649 and PP482, joined the programme. The increased weight of the Griffon 72 meant that a strengthened undercarriage was required. Other planned improvements included short-barrelled Hispano 20mm Mk 5 cannon and, importantly, a focus on surface finish with attention to panel fitting and eradicating surface irregularities. The importance of surface finish to performance had been established during the Second World War.

In this photograph of the first Mk 4 prototype, Z2118, the low profile of the leading-edge radiators is obvious. Z2118 first flew with the new engine and cooling system on 25 May 1945. When Z2118 was originally rebuilt with a Griffon 72, it retained the rounded wingtips of the Mk 1 and the triangular tailfin but later had clipped wings and the Mk 4-spec tail, as well as the definitive pointed spinner. The carburettor intake on Z2118 is the first version tried, extending further forwards and having a deeper, narrower mouth than the version seen in the top opposite image.

Opposite above: Third Mk 4 prototype MB649 in an interim specification somewhere between the original appearance of the Mk 4 and the definitive version. The aircraft differs from the Mk 1 in its engine and cooling installation – the wing-root extensions housing the leading-edge radiators are clearly apparent. The engine now powers a four-blade propeller. The tailfin has been enlarged and a dorsal fin fillet added to help offset the greater tendency to swing, imparted by the engine-propeller combination and to generally improve longitudinal stability. MB649 also demonstrates clipped wingtips to improve roll-rate and low-level speed. The carburettor intake beneath the engine cowling is the second of several iterations tried before a final version was settled upon. The propeller spinner is also an early version.

Opposite below: MB649 later in the Mk 4 trials process with a flame-damping exhaust. The Mk 4 programme always included the flame-damping exhaust as, towards the end of the Second World War, it was thought that a major part of the aircraft's role would be as a night fighter. In the event, the flame-damping exhaust did not progress onto production machines, and Fireflies were fitted with simple glare shields that protected the pilot's night vision but did nothing to cloak the exhaust flames from potential targets.

Above and below: Two photographs of Firefly Mk 4 prototype Z2118 in flight, taken by the renowned aerial photographer Charles E. Brown. By the time of these photographs, the Mk 4 had almost reached its definitive appearance. The wings and tail are as they would appear on all subsequent combat versions of the type, as are the propeller and spinner. However, the carburettor intake is the earliest version.

The prototype has now gained two permanent (but detachable) nacelles beneath the outer wings. It was accepted by this time that the Firefly was most useful in the reconnaissance role and as a fighter at night and in poor weather (when its performance deficit was less important). All aircraft would therefore be fitted with ASH radar. Various locations on the airframe were proposed but the best was on the outer wing, protruding forward of the leading edge. The inclusion of radiators in the wing root leading edges had displaced two fuel tanks, so a nacelle on the opposite wing to the radar housed a fuel tank as well as balancing the radar pod. (Radar was installed in the starboard wing nacelle, fuel in the port). In the photograph below, note the wing roundel wrapped around the nacelle. In later aircraft, the problem of how to apply the roundel with the nacelle being in the way was solved by simply moving the marking further outboard.

The Mk 4 was swiftly followed by the FR Mk 5. The latter was a 'universal' type incorporating the improvements of the Mk 4 but with the ability to convert quickly between the strike-fighter, anti-submarine and night fighter variants simply by exchanging equipment (and therefore gaining a different prefix: 'FR' for fighter/reconnaissance, 'AS' for anti-submarine and 'NF' for night fighter, though the latter was rare).

Above: A flight of seven production Firefly Mk 4s in company with four Seafire FR 47s, 1948–49. The photograph captures a period when FAA markings were in transition. The basic camouflage scheme was the same as during the Second World War, but the finish was now highly glossy to improve performance and to make the aircraft easier to clean. The Fireflies display a mix of national marking styles, most showing the later type introduced from 1947 with brighter colours and red, white and blue roundels in a 1:2:3 ratio and no fin flash. Some machines still have the older 3:4:8 ratio roundels with a narrow white ring. However, all have the later three-digit numerical codes starting with '2' to signify a two-seater aircraft.

Below: Firefly FR 1s from 1830 Squadron, Abbotsinch, aboard HMS *Illustrious*. This unit was originally a fighter/anti-submarine unit equipped with Fireflies and Seafires but standardised on Fireflies. From August to September 1949 it joined *Illustrious* for a period of embarkation and training. The squadron carried out 205 accident-free deck landings for which it won the annual Boyd Trophy.

The second aircraft back is wearing a 'D' tail code (indicating that it had been part of HMS *Illustrious*' ship flight until recently) and is equipped with rocket rails. (David Bull collection)

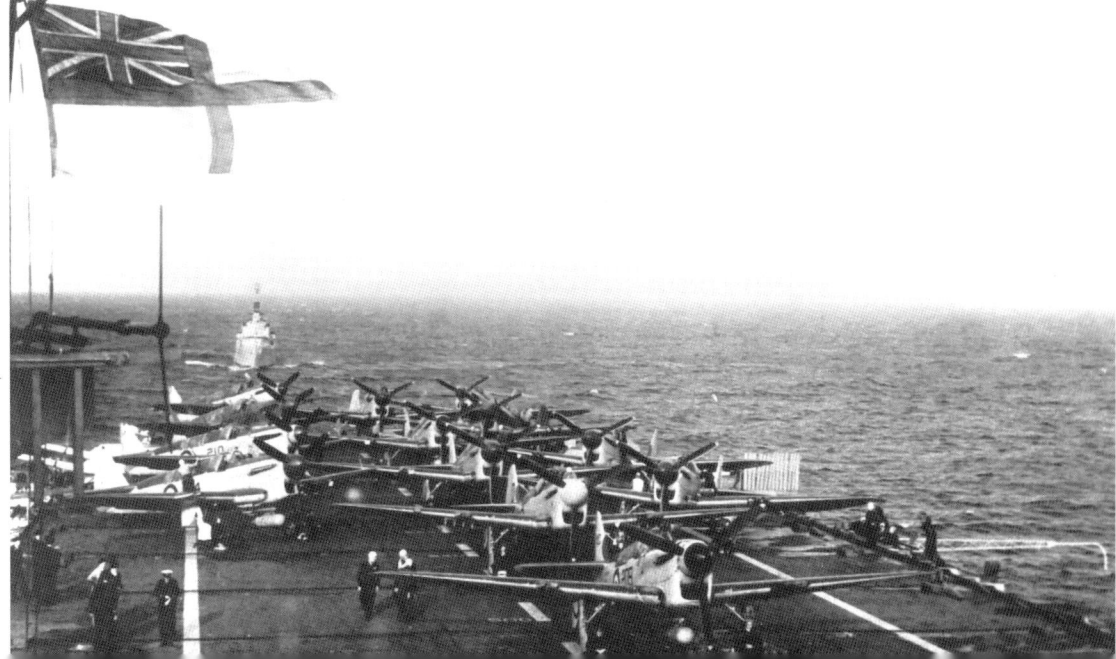

Above: Firefly FR 4 of 737 NAS, RNAS Eglinton in County Londonderry, Northern Ireland (indicated by the 'GN' tail codes) which re-formed in March 1949 as part of the 52nd Training Air Group. The squadron was equipped with Fireflies and Seafires to train aircrew in the use of air weapons and basic anti-submarine warfare. The Seafires left after a year, and from 1950 to 1955 the squadron used Fireflies exclusively (in a range of marks). Firefly 221 has caught a wire but swung to port and put a wheel over the edge of the deck. (David Bull collection)

Opposite above: Firefly 206 from 737 Squadron has suffered an undercarriage collapse and hit the barrier while moving sideways, and fire crews are preparing to tackle any resulting fire. However, Firefly Mk 4s had metal propellers, so they did not experience the lethal shattering of the blades as the wooden Mk 1 propellers did when they struck a barrier or the deck. Note how worn the paintwork on this once pristine aircraft is, following a hard life training aircrews. (David Bull collection)

Opposite below: Another deck crash in progress during training aboard HMS *Illustrious* in 1949. This time it is a Firefly Mk 5 (like the aircraft in above and opposite above photographs) based at RNAS Eglinton, but belonging to 719 Squadron, and training anti-submarine crews. Firefly 218 has hit the barrier so hard it has broken it and slithered on through, setting the deck crew fleeing for the safety of the catwalk. The impact was so severe it tore off the port wingtip. Intensive training was necessary to ensure mistakes and failures like this, if they had to take place, happened during a non-operational setting. By the time the crews took part in the Korean War from 1950 to 1953 (see Chapter 4), they were able to operate with astonishing levels of efficiency with very few accidents.

Post-war Fireflies

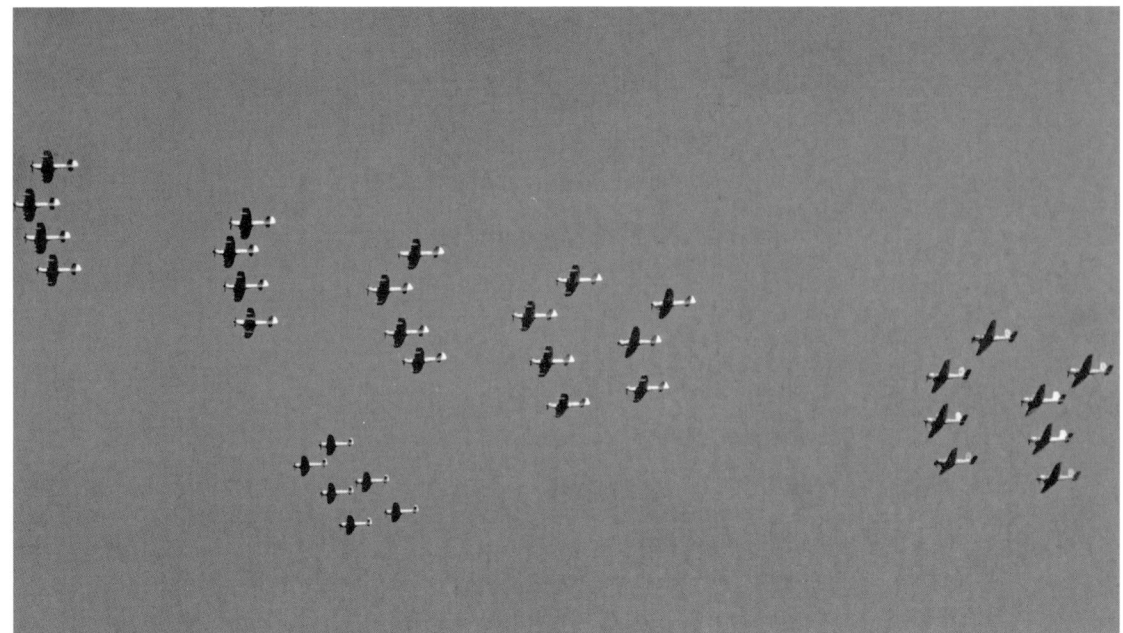

Above and below: Demonstrating the potency of the post-war FAA, aircraft from three fleet carriers: HMS *Implacable*, HMS *Vengeance* and HMS *Glory* stage a flypast in March 1950. The four 'finger four' formations and the 'vic' behind them from the left of the photograph are Fireflies. The two 'finger four' groups at the right are Blackburn Firebrands from *Implacable*, while the six aircraft to the lower left-middle, in two groups of three *en echelon* starboard, are Hawker Sea Furies.

A close-up of the Firefly formations demonstrates that two of the aircraft in the aftermost 'vic' are NF Mk 1s of a carrier's 'Black Flight'. The rest are Mk 5s.

Above: Firefly AS Mk 5 VT474/213 of 814 Squadron, HMS *Vengeance*, the carrier surrounded by pack ice during Operation *Rusty* on 15 February 1949 (the day the carrier's bow was damaged by an iceberg). Operation *Rusty* was an Arctic cruise to test the efficiency of carrier and aircraft in very cold conditions. On 9 February, the carrier entered the Arctic Circle and encountered pack ice on 15 February. A handwritten note on the reverse of this photograph reads, 'Here we are in the ice pack and this is where we got damaged. About 400 tons of water was leaking through right now'.

Firefly VT474 AS 5 later joined 812 Squadron (coded 811) and took part in the Coronation Review for Queen Elizabeth II.

The aircraft has slightly non-standard markings with a dark-coloured rudder and some of the nose panels wearing an older style of camouflage than the rest of the airframe.

Right: Firefly AS Mk 5 WB267 mid-crash aboard HMS *Vengeance* in September 1949 during the Home Fleet's autumn cruise. Flying operations were carried out off Scotland's east coast and Norway.

An ongoing problem with the Firefly Mk 5 was the bouncy nature of the undercarriage, which could pitch an aircraft over the wires if it touched down too hard, exacerbated by the under-damped arrestor hook being easily knocked back towards the fuselage. This appears to be what is happening here – the aircraft has bounced the hook clear of the wires, then landed heavily on its port undercarriage leg, which is collapsing as the Firefly heads inexorably toward the barrier.

Above: Firefly AS 5 VT365/027 of 703 Squadron based at Lee-on-Solent (as indicated by the 'LP' letters above the code) in 1948. The aircraft's markings are somewhat unusual and transitional. The short station letters were customarily carried on the tail (though after they were introduced, some aircraft wore them on the fuselage for a time). The nose panels were evidently replacements taken from a Mk 4 as they wear a completely different camouflage pattern from the rest of the aircraft. This Firefly later took part in the work-up of 817 Squadron of the Royal Australian Navy's Fleet Air Arm, and was assigned to the squadron during 1950. From 1952 to 1953, it was based in the Mediterranean and served with 810 Squadron. It was struck off charge on 26 March 1953. As seen here, it is carrying 3in RPs 'double stacked' with eight per side.

Below: A dramatic shot of a Firefly Mk 5 landing on HMS *Theseus* from flight deck level. After the Second World War, the FAA adopted the US style of carrier landing which involved bringing the aircraft in high and dropping it onto the deck. This exacerbated some of the early problems with the bouncy undercarriage oleos. Training and further development reduced accidents to a very low rate.

Above: Firefly AS Mk 5 VT393 ashore. This aircraft was delivered to the FAA on 13 February 1948 and had a relatively long career, being struck off charge on 31 January 1955. Here it is fitted with long-range auxiliary fuel tanks.

Below: A beautiful air-to-air image of a completely 'clean' Fairey Firefly AS 5, WB281, among the clouds. Unusually, the aircraft has no cannon or fairings, no radar or fuel nacelles, no national markings and no serial repeated under the wing. In this configuration, the elegance and cleanness of line of the Firefly 4/5 family is very apparent, despite requiring the space for a second crewmember and associated equipment.

WB281 was allocated to 810 Squadron at St Merryn on 12 Aug 1949. Its career lasted less than a year; on 8 Aug 1950, it floated over wires on HMS *Theseus* and collided with WB376 and WB369. Being badly damaged, it was effectively written off and was simply pushed over the side.

Above: Firefly AS Mk 6s (of 820 and 826 Squadrons) loaded with sonobuoys for detecting submarines, and Sea Furies aboard HMS *Theseus* in the Firth of Forth ready for Exercise *Mainbrace* (a large-scale Nato exercise off Norway) in September 1952. The UK government described the exercise thus:

> To test the forces of the Supreme Allied Commander-in-Chief, Atlantic, in co-operation with the forces of the Supreme Allied Commander, Europe, in defence of the northern flank of the North Atlantic Treaty Organisation area, and particularly to exercise the ships and aircraft of the N.A.T.O. countries taking part, in tactical co-ordination over an extended period.

The Firefly AS 6 reflected the increasing focus on anti-submarine warfare. The subtype could carry the full range of strike weapons as well as anti-submarine equipment, but the cannon were deleted.

Left: *Theseus* with Fireflies ranged aft during Exercise *Mainbrace* alongside the USN destroyer USS *Fred T Berry*.

Above: HMS *Eagle*, the newest and largest RN fleet carrier, which had only been accepted into service six months before, also took part in Exercise *Mainbrace*. Her air group pointed to the future, with 800 and 803 Squadrons' Supermarine Attacker jets alongside the Firefly AS 6s of 814 Squadron and Firebrands of 827 Squadron. She is seen here, probably during Mainbrace, with her entire air group minus the Firebrands ranged and an Attacker on the catapult ready to launch. Ironically, the Firefly outlasted the Attacker both aboard *Eagle* and within the FAA generally. The last Fireflies on *Eagle* were 826 Squadron's AS 6s in January 1955, while the last Attackers were those of 800 Squadron which left in May 1954.

Below: The fleet carrier HMS *Indomitable* in the Mediterranean, 1952–53. The carrier's Firefly squadrons (seen ranged aft) were 820 and 826 NAS with their AS Mk 6s. Both were on board when *Indomitable* suffered an explosion of her aviation fuel supply in February 1953 while the ship was carrying out flying operations. Nine crewmembers were killed and many aircraft were damaged.

Above: Fireflies of HMS *Indomitable* ranged ahead of the carrier's Sea Furies in 1953. Unusually, the tail letter is 'J' which does not correspond to the carrier's deck letter of 'A'. Here 12 Fireflies prepare to take off, armed with practice bombs in light-series stores carriers.

Below: This dynamic shot of a Firefly AS 6 catching a hook aboard *Indomitable*, 1952–53, is from an album assembled by a member of 826 Squadron labelled only 'The ill-fated 277'. Sadly, the serial of the aircraft is not identifiable and the ill fate is unclear.

Left: A wing of 16 Fireflies from 817 Squadron, HMAS *Sydney*, overfly a *Crown Colony*-class cruiser and an *Arethusa*-class cruiser as they take part in the Coronation Fleet Review of Queen Elizabeth II at Spithead on 15 June 1953. Seventeen full squadrons of Fireflies took part along with an individual Mk 7. The photograph was released to the press, with a special aircraft flying photographs from the review quickly to recipients (provided they were prepared to pay a premium) for the evening editions.

Below: Firefly F Mk 1 MB524 of 766 Squadron usually based at Lossiemouth (as indicated by the 'LM' tail letters). This aircraft took part with 766 NAS in the flypast for the 1953 Coronation Fleet Review flypast operating from Lee-on-Solent.

Right: An open day aboard HMS *Indomitable* in Casablanca in 1952. Fireflies are ranged along the port side of the flight deck.

Below: 1952 Fairy Firefly AS 6 WD913/275 of 826 Squadron, on HMS *Indomitable* in 1952, loaded up with a pair of 1,000lb bombs.

Above: HMS *Indomitable* and HMS *Theseus* at Gibraltar in 1952 with Hawker Sea Furys and Fairy Fireflies on deck. Note the cowling panels are removed on the Fireflies for maintenance showing the tubular steel engine bearers. The nearest Firefly is VT409 which also appears in the image on page 50 and parts of which survive in 2020.

Below: Firefly AS 5 VT409 before conversion to AS 6 as in the photograph above. VT409 was delivered on 9 April 1948 and withdrawn in September 1955. It was sold for scrap in March 1957, but the rear fuselage survived long enough for the late Nick Grace to acquire it as a source of spares for the restoration of WD833. The rear fuselage changed hands several times, appearing in at least two museums and is now with a private owner in Devon.

Above: Firefly 222 of 820 Squadron in the aftermath of a landing accident aboard HMS *Indomitable*. By now the aircraft displays the tail letter 'A' in common with *Indomitable*'s deck letter. Sadly, the serial of the aircraft is not readable and is partially painted over by the code.

Below: Pilot and petty officer crew chief sitting on a Firefly of 820 Squadron (note the 'flying fish' badge on the cowling) aboard HMS *Indomitable* in 1953. The Firefly is ranged at the extreme stern of the ship, and the round-down and deck letter 'A' is visible to the right. The Firefly is armed with sonobuoys for tracking submarines acoustically.

Above: Firefly T Mk 3 observer trainer PP523/264 (which had been built as an FR 1 but converted to Trainer status) of 796 Squadron. Based temporarily at Lee-on-Solent in 1953 from its usual home of St Merryn, PP523 took part in the Coronation Review flypast.

Below: Firefly AS 6 WD850 with the extensive range of stores that the aircraft could carry laid out. These include (from the outside in) RPs, mines, depth charges, supply cannisters, bombs, practice munitions, auxiliary fuel tanks and sonobuoys.

Above: Firefly TT (target tug) Mk 4 TW722, 1952–55. With the rapid withdrawal of the Firefly Mk 4 from front-line service, the remaining aircraft were transferred to second-line functions. In July 1951, TW722 was converted to tow targets for naval gunners to practice with. It spent most of its life as a target tug at RNAS Ford (as indicated by the 'FD' tail letters) with 771 and 700 Squadrons, though it did spend some time with the civilian-run Fleet Requirements Unit operated by Airwork at Hurn between 1957 and 1958. On 6 September 1954, the pilot became lost in bad weather, running low on fuel and suffering radio failure, so made a forced landing on Redcar racecourse. The aircraft was undamaged and after refuelling was able to take off from the racecourse without trouble (allegedly lifting off in two and a half furlongs). (Information from the Yorkshire Air Accidents website, see Bibliography.)

Left: An unusually painted Firefly operating as an 'Admiral's Barge' at Changi, Singapore, in the late 1950s. The aircraft the Firefly is taxiing past is a Hawker Sea Hawk F Mk 1 undertaking tropical weather trials. The Firefly is painted a single, overall dark colour (probably blue) with a Rear Admiral's flag painted on the cowling. Adding to the air of luxury, it even has white-wall tyres!

Chapter 3
Trainers

At the end of the Second World War, the production plans for military aircraft were reviewed. Any developments focused on the Pacific War were dropped immediately and longer-term priorities were examined. At the same time, certain programmes that might be considered a luxury in wartime could now be considered. The Admiralty raised some of these with Fairey including an advanced operational trainer.

Duncan Menzies, lead test pilot at Fairey Stockport and Fairey's liaison to the FAA, seized on that having seen first-hand how front-line aircraft had gained dramatically in power and performance, while training aircraft were generally similar to those in use at the beginning of the war. There was clearly a need for a training aircraft that filled the gap and allowed pilots to acclimatise to a much more powerful machine without having to jump from a 700hp 'advanced' trainer to fly solo in a 2,000hp single-seater, as well as a dual-control deck-landing trainer of similar performance to the latest service types. Menzies persuaded the Admiralty that the Firefly could be readily adapted to this purpose efficiently and cost-effectively, as the Admiralty already owned many now-obsolescent Firefly Mk 1s. He then persuaded the Fairey board to take it on as a private venture, as there would undoubtedly be interest from many air arms around the world.

Duncan's views were recorded by *Flight* magazine in January 1947:

> Mr Menzies confirmed on his travels that a relatively high percentage of mishaps occur as a result of incorrect handling, on or near the ground, by pupils unaccustomed to the 'feel' of modern high-performance fighters with relatively high wing-loadings. An average pupil may need four hours or more to familiarise himself with a modern fighter, and expensive accidents are not infrequent.

Below: Firefly F 1 Z2033, wearing 'class B' markings 'G-6-3' while acting as an aerodynamic mock-up for the Firefly Trainer

In the Firefly Trainer, the space formerly occupied by the observer's cockpit was filled by a second fully fitted-out pilot's cockpit, the floor 12in above that of the forward cockpit. This afforded the instructor an excellent view and required relatively little modification to the structure of the aircraft. A Firefly then at Stockport for repairs, MB750, was earmarked for conversion as the prototype, while Z2033 was mocked up with the revised rear fuselage shape for aerodynamic testing and first flew in that form in May 1946.

This photograph of the Trainer prototype wheeling in the clouds was taken by master aircraft photographer Charles E. Brown. The prototype retains two of the original four cannon so it could be used for armament training too.

As the Trainer was Duncan Menzies' 'baby', he made the initial flight in June 1946 even though he was no longer strictly a test pilot. Menzies then put on a two-day demonstration to senior officers of the FAA's Training Branch on 15 and 16 August, shortly before MB750 went to Boscombe Down for testing by the A&AEE. On 10 September, he flew it to Radlett for the Society of British Aircraft Manufacturers (SBAC) air show. Then in October, he toured the prototype around operational aerodromes giving service pilots a chance to try out the Trainer. In December 1946, the Trainer made a tour of France and the Netherlands.

Above and below: Two photographs of the Trainer prototype in late 1946. As an 'F 1', the prototype was part owned by Fairey, and a condition of its use by the Admiralty was that a Fairey test pilot always had to accompany service pilots. When the aircraft was being demonstrated at the School of Naval Warfare, the FAA instructors lost patience with the restrictions, leading to Duncan Menzies negotiating a solution between Fairey and the Admiralty for the aircraft to regain its service identity MB750. The paint shop at RNAS St Merryn added RN serials and roundels, and removed the 'class B' markings – but not the Fairey Aviation logo on the tail! MB750 was later sold to Thailand.

Left: Another cartoon from the mural at RNAS Dale by Sub-Lieutenant (A) 'Dax' Dashfield RNVR, 1947–48. This time the subject is very recognisable as the Firefly Trainer prototype distinguished by its black upper decking. It's unclear whether the incident depicted here of a near-miss with a Gloster Meteor jet fighter took place but, as other documented incidents appear on the mural, it is a possibility. The prototype is not thought to have visited Dale, but it did visit numerous RN air stations between 1946 and 1947 and Dax would have had plenty of opportunities to see it. (Photograph courtesy of Coastlands Local History Group and Tony Jupp)

Below: Firefly Mk 1s being rebuilt into Trainers at Fairey's assembly shop at Ringway, Manchester (the site for all final assembly and test flying for Fairey, Stockport). Although the Firefly was initially developed and built at Hayes, the Trainer was very much a Stockport project. Rebuilding a Firefly Mk 1 into a Trainer took around 1,500 man-hours – far less time than it took to build a new aircraft.

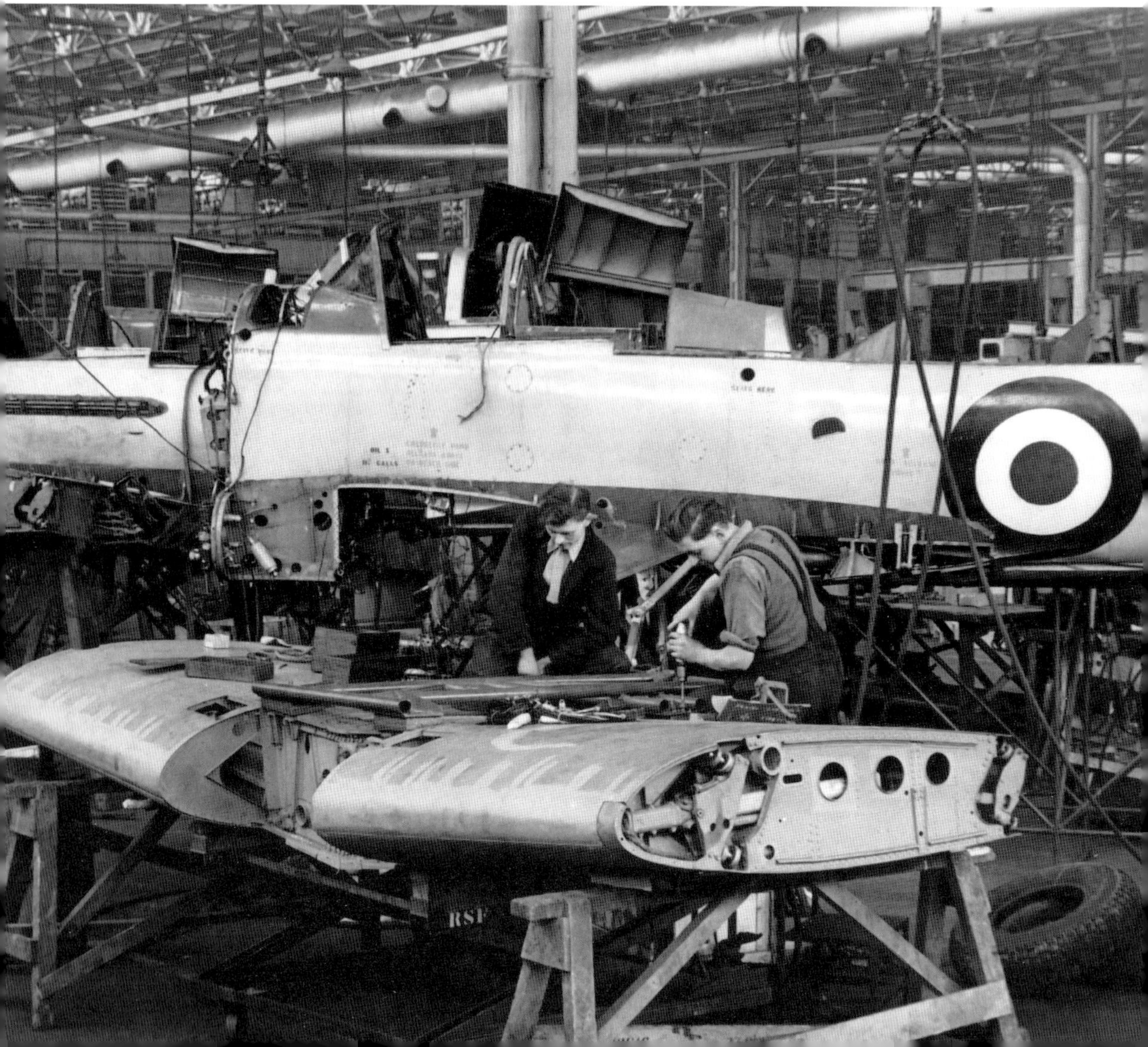

Above: A Charles E. Brown photograph of production Firefly Trainer Z2027 in flight. This is a T Mk 1 version which had no armament, the Admiralty initially considering that it was not necessary despite the prototype being armed.

Below: A Firefly Operational Trainer Mk 1, DV132, looking pristine just after leaving the factory in October 1947. There were 34 Mk 1 Trainer conversions in total.

Above and below: Two Firefly T Mk 2s, DK429 and MB721. The T Mk 2 was the next logical step; similar to the Mk 1 in many respects but, like the prototype, the Mk 2 was armed with two 20mm cannon and also had provision for a range of other weapons, such as 500lb bombs and other munitions. The T Mk 2 could act as an advanced, deck-landing capable pilot trainer (like the Mk 1), but it could also be used as a tactical weapons trainer. There were 57 T Mk 2s converted.

DK429 demonstrated the Firefly Trainer to industry and the public at the Society of British Aircraft Manufacturers display at Farnborough in 1948. MB721 is seen here at Ford, preparing to take off while Westland Wyverns approach the runway in the background.

Firefly Trainer T Mk 1 MB443/TG-X of the Royal Canadian Navy (RCN) in 1949. It was part of No.1 Training Group, Shearwater, attached to RCAF Dartmouth. The RCN took delivery of the *Colossus*-class light fleet carrier HMCS *Warrior* in 1946 and operated it with Firefly FR 1s (825 Squadron) and Seafire Mk 15s. Along with their purchase of front-line Fireflies, a number of Trainers were included: three Mk 1s and three Mk 2s.

Above and below: Firefly T 5 VX373, one of only four trainer versions of the Firefly Mk 5 to be built.

Although the Royal Australian Navy (RAN) did not initially order any dual-control Firefly Trainers, MB696 (a Mk 2 version of the aircraft with two 20mm cannon to add the armament trainer role) was used by 817 Squadron during its work-up at St Merryn in Cornwall. When the RAN began to operate its own Fireflies and Sea Furies in earnest, it quickly saw the benefit in having its own dual-control Fireflies. The modification to dual-control trainer status was engineered in such a way that it could be applied to existing aircraft. Fairey Stockport was therefore able to manufacture a kit for the RAN to convert aircraft locally to Trainer specification. Three kits were shipped to Australia, and Fairey Australasia produced at least one more kit in situ. The four aircraft known to have been converted between 1951 and 1955 were VT502, VT440, VX373 (which acted as prototype) and VX375. The Australian T Mk 5, as it was known, was the highest-performance Firefly Trainer developed and the only one that truly translated to the later operational variants.

The Royal Netherlands Navy (RNN) made similar modifications to some of its Firefly Mk 1s when they were replaced on the front line by newer models.

Chapter 4
Exports

Fairey pursued the possibilities for exports after the Second World War with vigour and was rewarded with a number of contracts for new or refurbished Fireflies. The first of these was for the Dutch Navy. This was, in fact, the first overseas aircraft sale for the British aircraft industry since the end of the Second World War. By the autumn of 1949, the type was being operated by three navies in addition to the RN. Fairey's in-house magazine proclaimed in September:

> WITH 4 NAVIES: A further batch of Fairey Firefly Mk 5 Fleet Reconnaissance Fighters has been accepted by the Royal Netherlands Navy. This new supply augments several previous deliveries of Firefly aircraft, which were the first post-war aircraft to equip the Royal Netherlands Navy. Pilots of the RNN were flown to White Waltham aerodrome, Maidenhead, to accept the aircraft and pilot them back to Holland. Recently, HMAS *Sydney*, Royal Australian Navy, and HMCS *Magnificent*, Royal Canadian Navy, departed British waters equipped with Fireflies, which are also in full supply to the Royal Navy.

Fairey went as far as to set up subsidiary companies in Australia and Canada to attend to the maintenance, overhaul and upgrade needs of the Fireflies and other naval aircraft in the service of those two countries.

Below: Three Firefly Mk 4s of the Koninglijke Marine (KM)(the Dutch name for the RNN) in flight. The first country other than the UK to operate the Fairey Firefly was the Netherlands. KM personnel formed 860 NAS during the Second World War and at the end of hostilities, this unit was transferred to Dutch control with the carrier HMS *Nairana* renamed HNMS *Karel Doorman*. The agreement for a supply of Firefly Mk 1s to assist the rebuilding of the KM was struck during 1945, with 30 aircraft to be delivered the following year.

A painting diagram was hastily prepared by Fairey with the new owner's markings in January 1946. However, in the event, the Mk 1s were simply delivered with British markings and the addition of an orange triangle on the fuselage.

The Firefly Mk 1s were in action in the East Indies again before long, as the Netherlands, trying to re-establish control of its colonial possessions, faced a rebellion. The Fireflies operated in support of marines on the ground. At the end of the year the KM ordered 40 FR 4s (as seen here), most of which ended up based in the East Indies. In 1952, they supplemented their stocks with AS 5s bought from Canada. (Dutch National Archives)

Left: Six Firefly FR 1s of 825 Squadron, Royal Canadian Navy (RCN), running up on HMCS *Warrior*, 1946–47. The first four are each carrying four practice bombs on one side only. The fifth aircraft has RPs fitted and the last appears to have long-range fuel tanks. This squadron was formally transferred from the RN to the RCN in January 1946 having built up for a year with increasing proportions of Canadian personnel. *Warrior* was commissioned into the RCN at the same time and arrived at Halifax in March. The carrier and its Fireflies and Seafires took part in fleet exercises in the Caribbean in the spring of 1947, from which this image probably dates. At the time, the Canadian Fireflies still wore their FAA markings and British roundels.

Below: Firefly FR 1 DK535 of the RCN resplendent in its newly applied livery of Dark Grey over Sky Grey, with maple leaf-shaped roundels. This colour scheme was unique to the RCN and unusual in Commonwealth operators of FAA types who usually retained the British colour schemes. Fairey worked hard to secure its deal to supply aircraft to the RCN. Part of this effort involved setting up a subsidiary to attend to their repair and overhaul: the Fairey Aviation Company of Canada, incorporated in 1948. According to the caption accompanying this Canadian Navy photograph, DK535 was 'The first Fairey Firefly to be repaired and overhauled by the Fairey Aviation Company of Canada Ltd, 1949'.

Right: Firefly FR 1 DK564 (post-overhaul and repainting into the new colour scheme) suffered this deck-landing accident aboard HMCS *Warrior* (1949–50). The RCN upgraded 825 Squadron to Firefly Mk 4s in 1949. The older aircraft passed to 826 Squadron before some time later being released for training, disposed of to other Firefly operators or returned to the FAA. Despite only using the Firefly Mk 1 for a relatively short period of time, the aircraft were well used, and the RCN operated more aircraft based on this subtype (26 FR 1s and six Mk 1/2 Trainers) than any other Firefly variant.

Below: Three Firefly Mk 4 aircraft used by 816 Squadron, Royal Australian Navy (RAN), during its training and work-up in the UK. They were based at RNAS Eglinton in Northern Ireland. The aircraft are TW730 (code 231), TW726 (228) and TW737 (227), all of which were returned to the RN when 816 Squadron departed for Australia. The squadron was commissioned at Eglinton in August 1948 and remained there training until it was time for them to embark in February 1949.

Above: Firefly FR 5 VT363, with radar and fuel nacelles conspicuously absent, carrying a pair of 1,000lb bombs. This machine was on RN charge, but in February 1949 it was used for catapult and arrester gear trials on HMAS *Sydney* to prove the carrier's equipment was functioning adequately for the new air group.

Left: In 1951, the delivery of the second RAN aircraft carrier, HMAS *Melbourne*, became subject to considerable delays, so HMS *Vengeance* was loaned to cover the intervening period. HMAS *Vengeance*, as she was named between 1952 and 1955, is seen here arriving at Sydney in March 1953 with a batch of 'embalmed' Firefly and Sea Fury aircraft on deck, welcomed by Bristol Sycamore helicopters. The RAN received 19 Fireflies aboard *Vengeance* to add to the 88 that had arrived at with HMAS *Sydney*.

Australian Navy Fairey Firefly AS 6 WB518/903 of 725 Squadron in 1968 after its conversion for fleet requirements duties, wearing the classic diagonal black stripes denoting a target tug. This aircraft was delivered to the RAN in 1950, transported on HMAS *Sydney*, and served with 816 and 817 Squadrons in the early 1950s. It was converted to tow targets in 1957 and served in that role well into the 1960s.

Firefly AS 6 WD901 was one of the 19 delivered to the RAN aboard HMAS *Vengeance* and served with 816 Squadron for a period until being relegated to second-line duties. It was converted to target-tug configuration by Fairey, Bankstown, as seen here. It was disposed of in 1966, being displayed in front of a scrap yard then restored to flight in Canada before crashing fatally in 1977 (see Chapter 8).

Below: Firefly Mk 5 WD827 was one of the second batch delivered aboard HMAS *Sydney* in 1950, for 817 Squadron. The aircraft suffered a wheels-up landing at Jarvis Bay in 1951 and was repaired at Fairey, Bankstown, before returning to flight the following year. Several more accidents followed and eventually WD827 became an instructional airframe, as seen here. This aircraft is currently preserved at the Australian National Air Museum.

Above, below and opposite above: Three photographs of privately run Fireflies from the Svensk Flygjanst AB (SFAB) company, Stockholm. Rather like Airwork in the UK, SFAB was a private contractor providing second-line services to the military. In 1948, SFAB was looking for a higher performance aircraft to replace its Miles Martinet target tugs. At the time, the Firefly had not been adapted as a target tug but was potentially suitable, with a strong, reliable airframe, a powerful engine and a rear cockpit with room for a winch and operator, not to mention space for a 7,000ft cable. It needed only a relatively short take-off and landing run, and would have a speed of more than 200 knots even when towing a sleeve drogue, glider target or flare.

In 1948, former FAA Firefly Mk 1s were released to Fairey to rebuild and sell. They had a windmill-driven Type B cable winch attached to the port side of the fuselage, forward of the rear cockpit, and various other modifications, such as cable guards and a small windshield to allow the operator to put his head outside the cockpit to visually inspect the tow. The SFAB Fireflies were all painted yellow overall, with black markings.

The first aircraft to be delivered was SE-BRA (the former DK568) in December 1948. A further 18 Fireflies were operated by SFAB until their withdrawal in 1964. A good proportion of the Mk 1s in current preservation owe their survival to SFAB, as they generously handed several to museums, including SE-BRD (previously Z2033 – the original Trainer mock-up which later became the target-tug prototype), as seen here. SE-BRL was the former DT939 and SE-BRF was formerly DT986.

Below: A representative of the Fairey Aviation Company formally hands over nine immaculate Firefly Mk 1s in desert camouflage to Lieutenant Colonel Asefa of the Ethiopian Air Force (EAF) at Ringway in 1951. The EAF wanted Fireflies due to the type's ruggedness, performance and strike capability. The purchase of the aircraft, though, was a protracted affair, beginning in 1948 but frequently held up by British vacillation and delayed when the Korean War began in 1950.

Eventually, with the entry of the Mk 6 into FAA service, enough older models could be freed up to supply some aircraft to Ethiopia, though fewer than they had wanted. Although the purchase of more Fireflies from Fairey was discussed, Ethiopia sourced the rest of its aircraft from Canada and the Netherlands when those countries began to move on from the Mk 1.

The Ethiopian Fireflies were mainly used for ground attack in Ethiopia's ongoing border conflict with Eritrea. They were used until the mid-1960s and then stored in flyable condition until 1969, when they were pushed out of their hangars and left outside to make room for newer aircraft. They slowly deteriorated and were thought to have been scrapped in the 1970s, but two were rediscovered in 1993 and transported to Canada where they are now being restored.

Left: Fairey Firefly Mk 1 SF11 at the Royal Thai Air Force (RTAF) Museum, Bangkok. The Royal Thai Navy purchased ten FR 1s and two T 2s in 1950, and received the first in June 1951. Unfortunately, between the deal being struck and the aircraft being received, the Navy had been involved in a failed coup and, as many of its officers had been detained, the Navy was not in a position to take on the Fireflies. The Air Force received the Fireflies instead, but did not really want them and quickly began running down the complement by refusing to order spares, instead cannibalising some of the Fireflies to keep others flying. They were operated until 1954. Perhaps surprisingly in view of the RTAF's indifference to the Firefly, one of them, J.4-11/94 (formerly MB410), was retained for the service's museum and was recently restored. (Cheryl Baumgärtner)

Below: The first Danish Firefly TT, 625 (formerly Z1842). The Royal Danish Air Force was another air arm to use the Firefly solely in its target-tug configuration. Initially, two aircraft, 625 and 626 (formerly Z2020), were acquired between October and November 1951. They were joined the following year by four more Mk 1s from Canada, with Fairey supplying conversion kits to alter the FR 1s to TT 1 standard. The Danish Fireflies were most often to be seen towing targets over the firing ranges on the coastal islands near Jutland until they were retired in 1959 and the three surviving aircraft sold to SFAB as a source of spares for its own Fireflies.

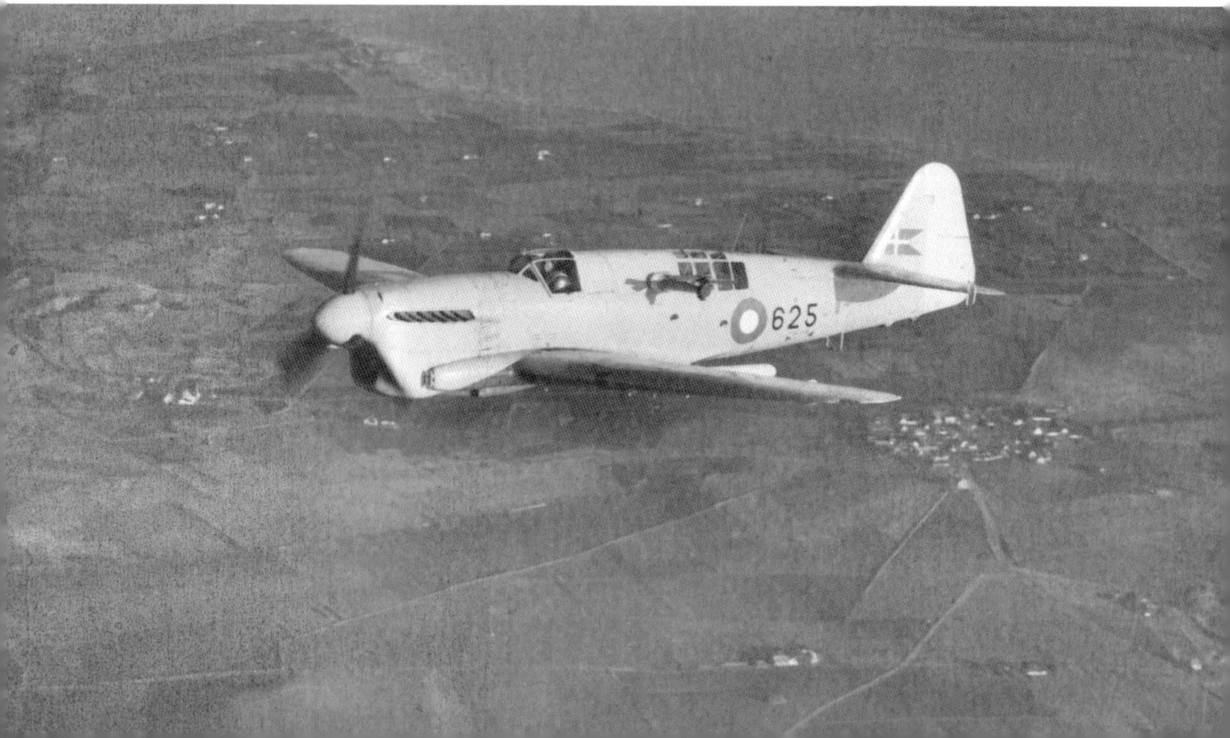

Above and below: Two air-to-air photographs of Indian Navy TT Mk 1 INS112 (previously DK477). In 1953, the Indian Navy identified a requirement for target tugs and considered the Firefly the most suitable, ordering five. Unlike most Firefly conversions, the Indian Navy's were not done at Ringway or the Stockport factory but by Fairey, Hamble, on the south coast of England. The five aircraft were numbered INS111–5, and they were joined by five Mk 4s in 1957, numbered INS116–20. They served until the 1960s. Large parts of one of the aircraft (identified as INS112 in some reports) survived. These were used as the basis of a somewhat rough restoration, and the aircraft marked as INS112 is currently on display at the Naval Aviation Museum in Goa.

Chapter 5
Korea

In April 1949, 827 Squadron (Firefly FR 1) embarked on HMS *Triumph* to assist the Far East Fleet with operations against communist rebels in Malaya. Operating from RNAS Sembawang, the squadron took part in Operation *Leo*, which involved aircraft co-operating with ground forces to progressively clear areas of jungle to force the rebels into a cordon. After the Malayan operations, *Triumph* took part in joint exercises with USN carriers which proved to be extremely fortuitous.

While *Triumph* was in the Far East, North Korean forces crossed the 38th Parallel into South Korea on 25 June 1950, triggering the start of the Korean War. The UN instituted a naval blockade, so *Triumph* provisioned at Kure, Hiroshima, and headed across the Korea Strait to act as the RN's main representative in the region until further forces could make their way out. Therefore, the Firefly was to enter its second major war.

Below: The light fleet carrier HMS *Triumph* putting to sea before her deployment to the Far East, sailing out of Malta's Grand Harbour for Sasebo in 1949, with the Firefly FR 1s of 827 Squadron ranged aft.

Triumph's air group was made up of the Firefly FR 1 and the Seafire FR 47, both older types that would only operate in the first months of the Korean conflict. Both aircraft took part in the first air strikes of the war, with nine Fireflies attacking the airfield at Haeju with 500lb bombs while the Seafires fired rockets. Then later in the day, the air group carried out strikes on railway targets. Throughout July, the Fireflies staged fighter patrols, photo reconnaissance and strike missions. The older types in use on *Triumph* were steadily depleted through accidents and breakdowns. However, the arrival of HMS *Unicorn* with additional aircraft and superior repair facilities helped both squadrons keep up with attrition. In September, *Triumph* sailed into the Sea of Japan for strikes on the port of Wonsan (an RN-only sortie) and operated in the Yellow Sea for the rest of the month enforcing the Western blockade. During this phase of *Triumph*'s operations, her aircraft were often engaged interdicting transport junks which North Korean forces had to use for supply as the coastal waters were too shallow and silted for ships. The Fireflys' ASH radar was particularly useful in these operations.

Above: A Firefly Mk 1 during a landing accident aboard HMS *Triumph* seen from the carrier's 'goofers' gallery'. On 29 August 1950, an 827 Squadron Firefly missed the wires and ran into the barrier. Tragically, the wooden propeller blades broke off and threw large fragments towards the island. The commanding officer of 800 Squadron, Cdr I. M. MacLachlan, was struck by a large piece of the propeller blade as he stood in the operations room. He died later of his injuries. This photograph shows a different incident but the danger from flying propeller blades can easily be seen – two of the three blades have broken off at this point in the incident and one can be seen high in the air towards the top left corner.

Right: A 'Black Flight' Firefly NF 1 (code 274) flying past HMS *Triumph* early in the deployment to Korea in 1950. The Firefly's hook and flaps are down, and the boom-mounted ASH radar pod can be seen very clearly with its slight nose-down attitude to improve scanning angles.

The text accompanying this British Official Photograph noted it was 'One of the first pictures of operations in which British and American carrier-borne aircraft were engaged recently in Korean waters'. The carrier in the background is USS *Valley Forge*.

Left: A well-known photograph of a Firefly FR 1 from 827 Squadron during a 'wave-off' on HMS *Triumph*'s deployment to Korea, while three more Fireflies enter the pattern to land. The black and white identification stripes that can be seen under the Firefly's wings, partially obscuring the serial, were added after a USAF B-29 shot down one of *Triumph*'s Seafires on 28 June 1950. Whereas, the large Union Jack on the flight deck was applied when *Triumph* returned to Sasebo to replenish.

Triumph's air group alternated between defensive flying for the entire task group, with Fireflies covering anti-submarine patrols and Seafires on combat air patrol, and strikes on shore targets, co-ordinating with USN carriers. In September 1950, *Triumph*'s aircraft supported the amphibious landings at Inchon with patrols and air strikes. Initially, losses to combat causes and accidents could be made good with deliveries of aircraft from the maintenance carrier HMS *Unicorn*, but by this time, the Far East Fleet (FEF) reserve had run out of the older types operated by *Triumph*, and her contribution was therefore at an end. She left Sasebo on 25 September, returned home via Malta, and 827 Squadron disbanded at RNAS Ford on 22 November 1950.

Below: A Firefly AS Mk 5 of 810 Squadron sinks towards the deck of HMS *Theseus* a moment after receiving the 'cut' signal from the batsman. *Theseus* replaced her sister ship, HMS *Triumph*, as the RN's 'duty' carrier in Korean waters in October 1950 after having been detached from the Home Fleet in August. In contrast to *Triumph*, *Theseus*' Firefly squadron now exclusively operated the latest variant, the AS 5, having dispensed with the last of her FR 4s after working up to combat capability.

Above: *Theseus* leaving Grand Harbour for the Suez Canal and the Far East in September 1950. This carrier had a significantly more up-to-date air group, consisting of 810 Squadron with its Firefly Mk 5s and 807 Squadron's Hawker Sea Fury FB 11s. The Firefly 5/Sea Fury 11 combination would prove to be the standard air group mix for the RN and RAN in Korea. From *Theseus* onwards, all Commonwealth carriers deploying to the region would operate these two types. For the smaller light fleet carriers, the combination offered the best mix of performance and hitting power. The Fireflies have had their Korea identification stripes added, while the Sea Furies have not. The Sea Furies are already armed with bombs, presumably in anticipation of a practice sortie.

On *Theseus* taking up station on 8 October 1950, in the Yellow Sea off the west coast of Korea, her air group was charged with reconnaissance and tactical missions in support of the advancing UN forces' left flank.

Right: Fireflies ranged aboard *Theseus* in preparation for a catapult take-off. The catapult was becoming an increasingly important feature of RN aircraft carriers due to the increasing size and weight of aircraft. The foremost aircraft is being guided to port off the centreline, to line up on the catapult track (the twin 'tramlines' that can be seen on the port side of the flight deck at the bow).

On 27 October, the steel wires that drew *Theseus*' catapult trolley were found to be in urgent need of replacement. Therefore, the carrier was restricted to launching aircraft by free take-off only and without bombs, rockets or drop tanks. Furthermore, six Fireflies had to be left behind in Japan to make the carrier's deck park more manageable. Naturally, this reduced the effectiveness of *Theseus*' air group, and with the UN seemingly on the brink of victory, the carrier withdrew for repairs. Remarkably, the engineering crew managed to renew the catapult wire at sea, which had not been done before, saving valuable time at the Hong Kong dockyard.

While *Theseus* was undergoing repairs, a massive Chinese assault caused a significant reversal, and UN forces were pushed back to the south. *Theseus* returned to the combat zone, now with the full number of aircraft again.

Left: This series of photographs (see pages 74–76) shows Firefly deck operations aboard HMS *Theseus*. This aircraft, WB334/231, was damaged on the voyage out to the Far East and was broken up for spares at Sembawang, before being written off on 22 September 1950. It was less than a year old, having been delivered on 3 October 1949.

In December 1950, *Theseus* was at sea for 23 days without a break, and her aircraft flew more than 600 sorties in that time. *Theseus*' aircraft successfully completed 1,463 carrier landings without an accident (and when this run was broken, it was a Sea Fury that broke it, not a Firefly). *Theseus*' rate of operations and overall efficiency was so great that the air group (the 17th CAG) was awarded the Boyd Trophy, which was given each year for the greatest feat in RN aviation. The following series of photographs give an impression of what a well-oiled machine both ship and air crews had become.

Below: A range of four Sea Furies and eight Fireflies await take-off. The aircraft are in the midst of starting their engines, with three of the four Sea Furies running, along with all but the aftermost two Fireflies on the starboard side of the flight deck.

Above: The Sea Furies have taken off and the Fireflies are beginning their take-off runs. Here one just begins to raise its tail while another Firefly waits behind.

Right: The last Firefly raises its tail on the throttle being opened and rolls forward. The amount of left rudder required to keep the aircraft running straight, against the torque of the Griffon 72 driving a four-bladed propeller, is apparent.

One of the Fireflies is now on its final approach to land aboard *Theseus*. The batsman guiding the aircraft onto the correct path can just be seen through the slats of the windbreak, signalling that the Firefly is on the right path.

Above: The Firefly has caught the second wire and come to a stop and, although the aircraft is somewhat off-centre, it is a good landing, and deck crew from the port catwalk are running to free the hook.

Left: The deck crew are now rolling the aircraft back to help free the wire from the hook.

A moment later and the aircraft taxies forward to 'Fly One' to park, so the barrier can be raised and the next aircraft can land. However, the arrestor wire has evidently caught on one of the other wires when reeling back in, and a deck crewman is running to free it.

Right and middle: A Firefly AS 5 following a barrier strike aboard HMS *Theseus*, 1949–50. These became increasingly rare as the combat-hardened carrier crews got their roles down to a fine art, maintaining an accident-free sortie rate that contemporary jets could only have dreamed of. The 17th CAG undertook 3,446 sorties while engaged in the Korean conflict. For the Fireflies, this was a mix of strike, reconnaissance (particularly railway marshalling yards) and anti-submarine patrols. After ten patrols, the last in company with USS *Bataan* off the east coast, *Theseus* returned home to be replaced by HMS *Glory*. Firefly WB334/231 of 820 Squadron was repaired and continued to serve with the FAA until 1955.

Divisions! HMS *Theseus* celebrates her successful deployment in Korean waters, entering Sasebo after leaving Korean waters for the last time with her air group; 11 Fireflies, 17 Sea Furies and crew, arrayed on deck in ceremonial order. Just visible on the island is the carrier's display of 'battle honour' scrolls, as seen more clearly in the middle image on this page. This shows that the 'Korea 1950–51' battle honour (the first earned by an HMS *Theseus* since the First World War) has already been added at the bottom of the display, clearly something the crew was justifiably proud of.

Above: Firefly FR 5 WB271/204 was close to becoming a veteran of the Korean War having been shipped to the theatre in 1951 as a replacement aircraft if required. Here she is seen in the markings of 204 of HMS *Glory*'s air group, 812 Squadron, which had operated the Firefly since just after the end of the Second World War (see Chapter 1). WB271 was operated and displayed by the Royal Navy Historic Flight (RNHF) for more than 30 years until its tragic loss in 2003.

Glory arrived in the Far East in April 1951 after a spell in the Mediterranean. Her aircraft commenced operations before the end of the month. The Fireflies and Sea Furies (of 804 Squadron) carried out a string of strike missions against transport infrastructure, such as railways and bridges, and junks off the coast. The incredible efficiency demonstrated by *Theseus* during her deployment was matched and even exceeded by *Glory*; the new carrier set a record of 89 sorties in a single day, and her CAG completed more than 1,000 deck landings without an accident. Between April and September, the carrier suffered only nine accidents.

WB271 was shipped to the Far East as deck cargo aboard HMS *Warrior* and was unloaded at Sembawang, Singapore, at the end of the year. As it turned out, WB271 was not required for Korean War service and, after the end of hostilities, was transferred to the RAN for service with 816 Squadron (HMAS *Sydney*'s air group).

Opposite above: The RAN carrier HMAS *Sydney* off the coast of Korea with Fireflies ranged aft alongside the Canadian destroyer HMCS *Athabaskan* in late 1951.

HMAS *Sydney* was the next carrier to represent the Commonwealth in the Korean War, replacing *Glory* in September 1951. The carrier had only arrived in Australia two years before, and it was an important statement of intent for the youthful Australian Fleet Air Arm to take on an important combat deployment. To say *Sydney*'s air group did the RAN proud would be an understatement. *Sydney*'s first patrol began on 3 October. Remarkably, her company reached such a state of efficiency that within a week she had equalled *Glory*'s record for sorties in a single day. Among the tasks the Fireflies of 817 Squadron undertook were close air support to the 1st Commonwealth Division and typical strikes on railway infrastructure.

The photograph was released on 13 November 1951, accompanied by the following text:

Carrier operations in Korean waters.

First official pictures of His Majesty's Australian Aircraft Carrier Sydney in action in Korean waters. The Australian carrier recently relieved HMS Glory. The Canadian destroyer 'Athabaskan' comes alongside while escorting HMAS Sydney during recent operations in Korean waters.

Right: Firefly casualty evacuation. An injured crewman on HMAS *Sydney* is helped into the observer's cockpit of an 817 Squadron Firefly FR 5 to be flown away for specialist treatment. The chief threat to life and limb for Firefly crews was anti-aircraft (AA) and small-arms fire from the ground during their ground attack missions. On 26 October 1951, in an incident reminiscent of the novel and film *The Bridges at Toko Ri*, Firefly WB393, flown by Sub-Lieutenant N. D. MacMillan, was hit by AA fire while attacking a railway tunnel and force-landed. MacMillan and his observer, CPO Hancox, took cover and held off North Korean troops with submachine guns carried for just such a scenario, with support given by *Sydney*'s Sea Furies, and later Royal Australian Air Force Meteors. Eventually, the carrier's rescue helicopter (loaned from the USN with its crew) was able to extract the two airmen and took them to the nearest UN airfield.

Above and below: Firefly WB316/203 during and after a landing accident aboard HMS *Unicorn* (the repair and maintenance carrier in 1952). *Sydney*'s air group was supplemented from aircraft in RN FEF stocks, which is where this aircraft came from in September 1951 along with three other Fireflies and several Sea Furies. The aircraft took part in numerous strikes and other missions (notably strikes against rail bridges), flown several times by Lieutenant Colin Champ. During one mission on 11 December, the aircraft was hit in the wing by ground fire, but Champ was able to return safely to *Sydney* and the aircraft was repaired. It was probably when WB316 was being returned to the RN at the end of *Sydney*'s deployment, transferring aboard *Unicorn*, that this accident happened. *Unicorn*, with her unique design and very high topsides, suffered from turbulence on the flight deck, which might have contributed to this incident. Despite damage to the undercarriage, flaps and propeller, the aircraft was evidently repairable as it returned to RN service, including aboard HMS *Vengeance* and later back to the RAN and HMAS *Sydney* in 1953. During *Sydney*'s time on station, her air group lost 15 aircraft in action, with three aircrew killed.

For more information on Lieutenant Colin Champ RAN's career during the Korean War, see 'An Australian Naval Pilot's Career' website, the Fleet Air Arm Association of Australia and the ADF-Serials websites (details in the bibliography).

Above: A full range of Fireflies and Sea Furies aboard HMS *Ocean* before a practice mission off Malta. Firefly FR 5 WB344/230 is in the front while a commander RN looks on from the 'goofers' gallery'. *Ocean* was serving with the Mediterranean Fleet before her call-up to replace *Sydney* in early 1952. Her first patrol was off the west coast of Korea, carrying out strike missions against a range of targets. In one day she flew an astonishing 123 sorties (47 of which were by 825 Squadron's Fireflies) and expended 90 tons of munitions. Note the deck crew lying by each mainwheel ready to pull the chocks away.

Below: Sea Furies and Fireflies on the deck of HMS *Ocean* in 1952. This carrier's period on station coincided with the arrival of MiG-15 jets in theatre. It was 802 Squadron Sea Furies from this carrier that became the first piston-powered aircraft of the war to shoot down a jet. As well as beating the already impressive sortie rate records of previous light fleet carriers, *Ocean* developed several tactical innovations, such as launching strike aircraft before dawn (a tactic designed to catch out troop and transport movements in the gloom before full daylight). The now standard attacks on railway bridges continued and, unsurprisingly, *Ocean*'s CAG excelled at these too, with eight bridges knocked out in two days. The likelihood of 'friendly fire' may well have diminished by this time as the various UN services had become more familiar with each other's aircraft. Photographs from this period suggest not all *Ocean*'s aircraft wore the black and white identification stripes, at least not at all times – though some certainly did.

Four bomb-armed Fireflies line up to take off from *Ocean* in 1952 while another four aircraft pass by overhead waiting to formate. On 16 September 1952, *Ocean*'s aircraft destroyed the last remaining railway bridge between Chinnampo and Pyongyang.

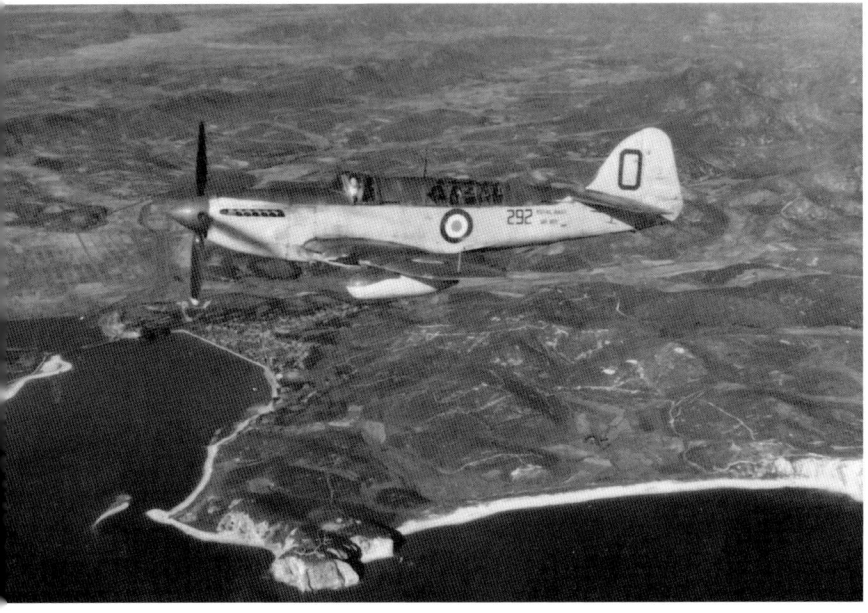

Fairey Firefly WT405/292 from 825 Squadron, HMS *Ocean*, flies a reconnaissance over Korea's eastern coast in 1952. For the 1952 Korean deployment, *Ocean*'s CAG was awarded the Boyd Trophy.

A Firefly FR 5 aborting a landing on HMS *Ocean*, which is off Korea, in 1952. The Firefly seems so low it was in danger of colliding with the round-down which could pull out the hook and lead to a barrier strike. The batsman has jumped into his 'pit' as the aircraft drifts over his head with relatively little clearance.

Above: Fireflies and Sea Furies aboard HMS *Glory* in 1952 during her second Korean deployment. The Fireflies are fitted with RPs, which by this time were available with a special 'shaped charge' warhead for targets like bunkers. On 5 April 1953 (a day of intense operations), *Glory* equalled *Ocean*'s record of 123 sorties in a day which meant every regular pilot flying four times, and occasional pilots such as the Commander (Air) and Flight Deck Officer flying at least twice. *Glory* spent two years away from the UK and during two spells in Korean waters, her aircraft accounted for 70 bridges, nearly 400 road vehicles and around 50 units of railway rolling stock. She handed over her duties, as well as spare equipment and stores, to *Ocean* in May 1953 before returning home.

Below: The badly damaged tail of a Firefly from HMS *Ocean*'s CAG. *Ocean* had taken on a new air group after withdrawing from Korean waters in October 1952 and now included the Fireflies of 810 NAS. Since *Ocean*'s first spell on station, North Korean forces had developed means of countering the dawn raids her air group had had so much success with. They used crude but effective means of directing fighters onto the raiding aircraft, so the tactic was riskier than it had been previously. This Firefly was lucky to retain enough control to return to the carrier with a fire in its tail that burned off all the rudder fabric and might have burned through the control cables. Fortunately in this case, the Firefly's elevators were metal skinned having changed from fabric skinning early in development.

Ocean's CAG carried out more close air support for the Commonwealth Divisions and later disembarked three Fireflies to Pyongtaek to act as night fighters, as aircraft such as Polikarpov Po-2 biplanes were carrying out raids under cover of darkness but were too slow for jet night fighters to engage. An armistice was signed on 27 July, but *Ocean* remained in theatre until October to police the ceasefire.

Chapter 6
Mk 7s and Drones

In the late 1940s, the Admiralty was preparing for the dedicated anti-submarine Fairey Gannet, but this was likely to take many years to bring into service. An aircraft with greater capabilities than the existing Firefly AS 6 was required in the meantime, particularly one with space for a third crewmember and greater load-carrying ability and endurance. High performance was less important. Fairey agreed that a modification of the Firefly would be possible, with an enlarged rear cockpit with a bulged canopy, subject to aerodynamic tests on a flying mock-up as with the Trainer. This variant was given the designation AS Mk 7.

Remotely controlled aerial targets had been developed before the Second World War, but developments in anti-aircraft weapons, particularly air-to-air and surface-to-air missiles, in the late 1940s and early 1950s meant more sophisticated and higher performance targets were needed. The Royal Aircraft Establishment began developing the Firefly in this role in 1952 with assistance from Fairey (which had established a division to develop air-to-air missiles and pilotless research vehicles). The first successful use of a Firefly drone 'in anger' took place in September 1955 when a U Mk 8 (the drone version of the Firefly Mk 7) was shot down over Aberporth.

Below: Firefly Mk 7 aerodynamic testbed MB757 (originally built as a Mk 1) pictured in October 1949. The external profile of the new observer's cockpit was added in a solid form. The tail was a standard Mk 4/5/6 form, while the additional weight aft was balanced by rearranging the engine cooling configuration. A layout similar to that of the abortive Mk 3 was adopted with the large beard radiator adding weight far forward. The rounded wingtips of the Firefly Mk 1 were also reintroduced to increase wing area and therefore reduce wing loading.

MB757 after the mock-up rear cockpit had been replaced with the real thing. The radome of the starboard wing nacelle has been removed, which helpfully shows the extended wing root that represented an effort to increase wing area.

Right: The prototype AS Mk 7 proper, WJ215 (which differed visibly from WB757), with a heavily baffled exhaust shroud and a more bulged cockpit canopy hood than the standard item. It was also fitted with longer travel oleo legs and a longer span tailplane. WB215 first flew on 22 May 1951, which is when this image dates from.

Right below: Fairy Firefly AS 7 WJ146 at RNAS Lee-on-Solent. This aircraft first flew on 16 October 1951 but was not delivered until 25 July 1952. At the time WJ146 left the factory, the AS 7 programme had reached crisis point with myriad changes needed to make the aircraft fit for service use. The tailplane would have to be completely redesigned. The larger radar nacelle, seen here prominently, was bigger than that of the Mk 4/5/6 to house the larger ASV 19A radar and was causing instability. It was highly unlikely that the aircraft would ever be suitable for operation from an aircraft carrier, especially at night.

Above: Firefly Mk 7 at the SBAC show at Farnborough in 1953. By this time, the efforts to make a workable anti-submarine aircraft of the Mk 7 had failed, despite the significant changes made to the basic Firefly, such as the huge, tall tailfin and rudder (introduced to address the considerable erosion in directional stability). Despite valiant efforts by Fairey to improve control, and some movement in the right direction, the AS 7 was cancelled on 23 April 1952. Some were repurposed as observer trainers, which had no need to operate from carriers and benefitted from having one more place in the rear cockpit for pupils than the T 3. These were re-designated T 7. Others were adapted into uncrewed target drones and designated U 8.

Below: Firefly Mk 7s in formation. These aircraft were built as anti-submarine AS Mk 7s but were only used from shore bases to train observers. The aircraft include WJ192/302, WJ168/300 and WM765/303, which were all from small pre-production batches. The 'MF' tail codes indicate that the aircraft were based at St Merryn from 1951 to 1953.

Above and right: Two photographs of Firefly U Mk 9 WB257 (an example of the drone conversion of the Mk 5). A second batch of drone Fireflies was ordered in 1955 after the Mk 7 had stopped production, so a version of the Mk 5 was introduced. The wingtip pods carry recording cameras.

Beneath the wing in the image on the right is an example of the Fairey Fireflash air-to-air missile which was tested at the Llanbedr range using Firefly drones.

Right and below: Two photographs of U 9 VT413 which was built as an FR 5 in 1948. In addition to the sites associated with testing with drones in the UK, a facility was established at Hal Far, Malta, with 728B Squadron operating the drones. It was here that VT413 ended its days, shot down by a Sidewinder launched by a Supermarine Scimitar in 1961.

Chapter 7
The Firefly Close-up

The engine and radiator installation from the Firefly Mk 1. Note the twin circular carburettor intakes beneath the rear of the engine. These were fed with air taken through ducts running along the side of the 'beard' radiator housing.

The distinctive 'beard' radiator housing of the Firefly Mk 1 (an NF Mk 1 in this case but all sub-variants were the same). The sectioned-off area on the right of the radiator (left on the photograph) is the oil cooler, while the rest is the coolant radiator. Part of the flap to increase the flow of cooling air can be seen at the bottom of the picture. Shown to good effect here is the split carburettor intake, with ducts extending forward from the wing leading edge to terminate in the 'cheek' position. Note the mesh over the intake to prevent foreign object ingestion.

Close-up view of the port side carburettor intake on the Mk 1.

Right and below: The Mk 1's pilot's cockpit shown with canopy closed and open. The cockpit windscreen and hood seen here are of the revised version introduced from the 71st Mk 1 to improve pilot view, headroom and to try to prevent the canopy detaching. Later, a one-piece 'blown' hood eliminated all framing.

A Firefly Mk 1 aft cockpit also showing some of the wing stowage mechanism.

A Firefly Mk 1 fuselage from aft showing both cockpits. The distinctive 'hump' in the rear fuselage to help accommodate the observer and his equipment can be seen very clearly in this view.

The wing-fold mechanism seen on an early F Mk 1. The method of twisting the wing about a pivot in the rear spar as it folded backwards was drawn from Blackburn practice. It was both simple and compact. All aircraft up to the Mk 5 had manual wing folding, at which point hydraulic folding was finally incorporated.

Right: Detail of the landing light just inboard of the cannon barrels.

Below: The Youngman Flap, demonstrated on WB271, in the 'cruise' setting. The Youngman Flap was a version of a trailing-edge flap used to increase lift at reduced speeds. When H. E. Chaplin was working on the Firefly design, the Barracuda torpedo bomber had already been designed with a simpler form of the flap. The Youngman Flap was somewhat more complex than most contemporary flap designs, with a variable angle and leading-edge position, and a slot between the wing surface and the flap's upper edge to accelerate air over the aerofoil. The Firefly's Youngman Flap also retracted neatly into the lower surface of the wing, unlike that of the Barracuda which remained aft of the wing at all times. This made for a more complex mechanism, but it was well worth it for the improvements in fuel economy and manoeuvrability it generated, as well as helping a great deal with carrier landing and take-off.

Another photograph of VT365 (see Chapter 2) showing the flaps extended into the 'cruise' setting as well as a heavy load-out of 16 RPs in 'double-decker' mounts.

Close-up of a Mk 4 wing showing the long-range fuel tank, ASH radar pod and cannon fairings (for the later, short barrel Mk 5 Hispano 20mm cannon.)

Below left: Firefly Mk 5 bomb beam, 1,000lb mine, deflector chute and four American-type sonobuoys. The photograph is dated 8 June 1949.

Below right: A Firefly TT Mk 4, '501', with its Type G Mk 3 winch at the SBAC show, Farnborough, in 1953. The Mk 4 used this much more flexible external winch, rather than the internal fitting of the TT Mk 1. It meant that any Mk 4–6 could be converted easily to the target-tug configuration. This image also shows the leading-edge radiator housings clearly.

Chapter 8
Preservation

Fairey Fireflies have fared reasonably well in preservation, helped by their extensive use in second-line roles after their first-line career came to an end. Most, if not all, preserved Fireflies spent at least some time in second-line roles, and some worked for far longer as target tugs or trainers than they ever did as strike-fighters. There are numerous surviving aircraft in restored or unrestored condition, including a rare two-seater trainer and several Mk 1s. Thailand's Royal Air Force Museum has a recently restored Mk 1, as does the Indian Naval Aviation Museum at Goa. Canada and Australia both have several examples on display, the former including one in flying condition (WH632) and two former Ethiopian Mk 1s as well as several later marks. Further south on the North American continent is the second flyable example, WB518.

Below: Firefly TT 1 Z2033 at Staverton in the 1970s when it was owned by the Skyfame museum.

The biggest early boon to Firefly preservation was SFAB, which recognised the value of its Mk 1 Fireflies when it retired them in 1964 and passed a number to museums. This included Z2033, the early Firefly F 1, which was later the flying aerodynamic mock-up for the Firefly Trainer and then the prototype target tug. Z2033 was presented to the Skyfame museum at Staverton, where it carried out engine runs and wing folding demonstrations, wearing a generic Second World War camouflage approximating a Second World War FAA machine and the name 'Sir Richard Fairey' on the cowling. When Skyfame closed, Z2033 was acquired by the Imperial War Museum, Duxford, and restored to a more authentic condition, being given the markings of DK431 'Evelyn Tentions' from the British Pacific Fleet during 1945 (see Chapter 1), but retaining its correct serial. Following restoration, Z2033 passed to the FAA Museum, Yeovilton.

Fairey Firefly FR/TT 4 VH127 at RNAS Culdrose on 28 July 1971. This aircraft was a target tug with an FAA Fleet Requirements Unit, where it was identified as worthy of preservation. At the time it was accompanied by a Mk 7 which was also listed as a possible aircraft to be preserved, but presumably the RN decided it would rather not be reminded. VH127 is now preserved at the FAA Museum, Yeovilton, where it tends to alternate on display with the Firefly Mk 1 Z2033.

One of the 12 Fireflies purchased by the Royal Thai Navy, which went on to serve with the RTAF between 1950 and 1954. This example, J.4-11/94, is preserved at the RTAF Museum having been restored in the last ten years. Prior to its Thai career it was the FAA's MB410, serving with 1791 Squadron until an undercarriage collapse sent it back to Fairey for repair. While there, it was released for export. (Cheryl Baumgärtner)

Below: Representative of the RCN AS Mk 5 was this former Australian aircraft, WD901 CF-BDH (see Chapter 4), which was acquired by Canadian Warplane Heritage after Canada retired the type without retaining any for preservation. The RCN operated 18 AS 5s from 1949.

It is seen here in the 1960s. WD901 tragically crashed into Lake Ontario on 2 September 1977, leading to the death of the pilot and the loss of the aircraft.

Right: WB271 peeling away from its RNHF stablemate, Fairey Swordfish LS326. This image provides another demonstration of the utility of the Fairey Youngman flaps for slow flight – the cruise setting effectively makes the Firefly a biplane.

Firefly FR 5 WB271 was spotted in a scrapyard in Australia by personnel from HMS *Victorious*, and the wardroom had a whip-round to save it from the scrapman's torch. It was presented to the FAA Museum and was later identified as suitable for restoration back to flight. It joined the RNHF and served for many years as part of the RN's flying commemoration of FAA history. (Menzies family collection)

Below: An official FAA photograph of three aircraft of the RNHF, WB271, LS326 and Sea Fury TF956, saluting the last of the RN's conventional aircraft carriers, *Ark Royal*, in 1974. Sadly, both monoplanes have since been lost to accidents (WB271 with the loss of the crew in a crash at Duxford in 2003). The photograph was printed in the *Daily Telegraph* with the following short and rather inaccurate story:

A Swordfish torpedo bomber (centre), the last one still flying, and two veterans of the Korean war, a Sea Fury fighter and a Firefly strike aircraft, passing over the carrier Ark Royal, 50,000 tons, in Plymouth Sound to commemorate the Diamond jubilee of the Royal Naval Air Service. The aircraft are from the Fleet Air Arm Museum at Yeovilton, Somerset.

Since the loss of WB271, the UK has not had a flying Firefly, though it is hoped that might change with the restoration of another Svensk Flygtjanst Mk 1 at Duxford.

Select Bibliography

Brown, D., *Carrier Fighters* (Macdonald and Jane's, 1975)
Brown, E., *Wings of the Navy* (Airlife, 1987)
Brown, J. D., *Carrier Operations in World War II, Volume 1* (Ian Allan, 1968)
Bussy, G., *Fairey Firefly: Warpaint Series No. 28* (Hall Park Books, 2000)
Harrison, H., *Fairey Firefly: The Operational Record* (Airlife, 1992)
Hobbs, D., *The British Carrier Strike Fleet after 1945* (Seaforth Publishing, 2018)
Jones, B., *The Fleet Air Arm in the Second World War, Volume 1, 1939–1941* (Navy Records Society, 2012)
Jones, B., *The Fleet Air Arm in the Second World War, Volume 2, 1942–1943* (Navy Records Society, 2018)
Sturtivant, R. and Ballance, T., *The Squadrons of the Fleet Air Arm* (Air-Britain, 1994)
Taylor, H. A., *Fairey Aircraft since 1915* (Putnam, 1974)
Thetford, O., *British Naval Aircraft since 1912* (Putnam, 1991)

Websites

'An Australian Naval Pilot's Career', www.devboats.co.uk/ceclogbook/cec0715.php
'Fleet Air Arm Association of Australia', https://www.faaaa.asn.au/fairey-firefly-aircraft-histories/
'ADF Serials', http://www.adf-gallery.com.au/
'HMS Vengeance', www.hms-vengeance.co.uk
'Hampshire Airfields' (Lee-on-Solent page), http://www.hampshireairfields.co.uk/nos11.html
'Aircraft Accidents In Yorkshire', http://www.yorkshire-aircraft.co.uk/
'Armoured Carriers', https://www.armouredcarriers.com/